Vision in Japanese entrepreneurship

Vision in Japanese entrepreneurship

The evolution of a security enterprise

H. T. Shimazaki

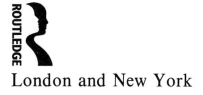

London and New York

First published 1992
by Routledge
11 New Fetter Lane, London EC4P 4EE

Simultaneously published in the USA and Canada
by Routledge
a division of Routledge, Chapman and Hall, Inc.
29 West 35th Street, New York, NY 10001

© 1992 H. T. Shimazaki

Typeset in 10/12pt Times by Witwell Ltd, Lord Street, Southport
Printed and bound in Great Britain by
Biddles Ltd, Guildford and King's Lynn

British Library Cataloguing in Publication Data
A catalogue record for this book is available from the British Library.
ISBN 0-415-08357-5

Library of Congress Cataloging-in-Publication Data
A catalogue record for this book is available from the Library of Congress.
ISBN 0-415-08357-5

Contents

List of illustrations   viii
Acknowledgements   ix

1   **Introduction**   1
2   **Early years**   8
    The setting   8
    School life   12
    Family business   18

3   **Business experience**   22
    Life as an employee   22
    Friend and partner   26
    Deciding on a venture   29

4   **Originating an industry**   38
    Preparation   38
    Initial struggle   46
    Tokyo Olympics   53
    Growth   57
    'Fame'   63

5   **Technology-based innovation**   68
    Remote alarm system   68
    Accusation of espionage   76
    Internal theft   77
    The alarm system proves itself   81

6   **Internal and societal adjustments**   86
    Managerial expansion   86
    On-site security   91
    Abolition of periodic security checks   94
    Legal controls   98

7 Cultivation of human resources   108
   Stock exchange listing   108
   Employee training   111
   The Human Development Centre   114
   Study scholarships   118
   Chief Executive Officer   121

8 Innovative ventures   123
   Knowledge exchange   123
   High technology at home base   124
   Expanding frontier   126
   Fire extinguishing system   128
   Entry control   132
   Nuclear security   134
   Household security   136
   Total building service   139

9 Vertical integration   143
   Production consolidation   143
   Corporate nerve centre   145
   SECOM   146

10 Internationalization   152
   Foreign expansion   152
   Australia   155
   Saudi Arabia   156
   Taiwan   158
   South Korea   169
   The United States of America   176
   Thailand   191

11 Research – foundation for growth   198
   Research consolidation   198
   Long-range research   205

12 Making history   214
   Data base service   214
   VAN service   220
   Computer security   222
   Community-based service   225
   New horizons   227
   Family front   231

13 Conclusion   233

Appendix: A social scientist's summary of entrepreneurship   242

Notes   244
Index   258

Illustrations

1 The Iida family, 1942. Makoto Iida (far left) with his
  parents (seated) and his four brothers.   11
2 Toda and Iida (back row left) with a Swedish security
  representative and Nihon Keibi Hosho's staff in front of a
  patrol car, 1962.   45
3 Nihon Keibi Hosho's guards arriving at the site of the
  Tokyo Olympic Village, 1963.   55
4 Iida (far right) and Toda (back row left) with the cast of
  the TV drama 'The Guardman', 1965.   64
5 Nihon Keibi Hosho's 'armoured car', 1966.   66
6 Control Centre, Tokyo, 1967.   75
7 Control Centre, Tokyo, 1978.   110
8 Iida (centre) and Hashimoto (far right standing) with
  SECOM's rugby team at its inauguration, 1985.   115
9 SECOM's Technical Centre, Tokyo, completed in 1986.   200
10 Iida receiving the Mainichi newspaper industrialist
   award, 1982.   240

Acknowledgements

The data on which this study is based was amassed from the Japanese and English language literature, personal correspondence, and interviews conducted from 1988 through 1991. The bulk of the Japanese data was collected during the first half of 1989 when I was on sabbatical in Japan.

I could not have completed this project without the generous help and cooperation of Makoto Iida, SECOM's founder and Chief Executive Officer, particularly in giving me access to corporate documents. Mr Iida does not like to look back. The only time he reflects on the past, he says, is on the anniversary of the company's founding. Over the past three years I have asked him to think and talk about the past a great deal. He has graciously accommodated me, for which I sincerely thank him. Quotations in the text attributed to Makoto Iida are excerpts either from taped conversations with him or from his personal statements which are included in corporate documents. This is true as well for quotations from other SECOM personnel.

I would like to express my appreciation to the members of SECOM's public relations office, especially its head, Zenjiro Kato, and Minoru Yasuda, for their assistance with this project from beginning to end particularly in data collection and field trip arrangement. I am indebted too to all the personnel from SECOM and its affiliated companies with whom I spoke in Japan, Korea, Taiwan, and the United States. They helped me to paint a broader picture of the corporation and the man at its helm. Although Mr Iida preferred that his private life remain just that – private – a few of his family members were willing to let me catch a glimpse of Makoto Iida, the man, and to them I express my thanks.

There are many others to whom I am indebted for their time and the information they shared with me. I would particularly like to mention

x *Acknowledgements*

Haruo Nishimura, former chairman of the department of crime prevention and juvenile studies at the National Research Institute of Police Science, and Dr Osamu Asano, physician and writer.

Editorial assistance was graciously provided by Aldo Opel, Alberta Research Council; Jim Langston, QC, Chief Crown Prosecutor; and Geoff England, professor, Faculty of Management, University of Lethbridge. Bruce Freeman assisted in both a research and editorial capacity as the project was nearing completion. Hitoshi Wada of SECOM's International Division carefully perused the manuscript to ensure it's accuracy.

I am grateful for both the moral and practical support received throughout this undertaking from the Faculty of Management at the University of Lethbridge. George Lermer, the Dean, has been considerate of my time requirements. Karen Eriksen and Holli Tamura have graciously complied with my numerous requests to type and retype the manuscript. The support and help extended to me by Beverly, my wife, is appreciated.

1 Introduction

This book focuses on Makoto Iida, Japanese entrepreneur, and
SECOM, the company he founded, Japan's first and, today, largest
security corporation. A study of one illustration of Japanese entrepreneurship, it also encompasses elements of a biography of Makoto Iida
and a business history of his company.

Various influences were at play in my selection of SECOM as the
subject of my present study. SECOM forms part of the service sector
industry. While some investigation has been carried out on entrepreneurial endeavour in the manufacturing sector in Japan – studies on
Soichiro Honda, Akio Morita of Sony, Tadao Yoshida of YKK,
Kazuo Inamori of Kyocera, Takeshi Mitarai of Canon, and Konosuke
Matsushita of Panasonic come to mind – in the English language
literature little has been reported on the service sector.[1] This oversight
is surprising given the growing importance of the service sector. In
recent years there has been a reversal in the relative size of the primary
and tertiary industries with the service sector now dominating the
economic scene. In 1990 *The Economist* reported that when the
agricultural sector is excluded, of Japan's 6.5 million businesses, more
than 5.5 million are in services and other tertiary fields.[2] Led by Japan,
the service sector in the Pacific region is rapidly becoming a new
foundation for economic development destined to have a profound
global impact.[3]

Having determined to study entrepreneurship within the service
sector, I chose to focus on SECOM, a comprehensive security/
communications firm. Japan is recognized both domestically and
abroad as one of the world's safest industrialized nations. SECOM's
phenomenal success in marketing security measures in a country
where there was little perceived need for them before their introduction was a strong indication of entrepreneurial processes at work. My
interest as a cultural geographer in people's creative responses to their

surroundings and events triggered a desire to learn about the person or people behind these processes.

The phenomenal success of certain Japanese companies has captured the attention of business leaders around the world. As interest in Japanese management has widened, English language publications about Japanese business have multiplied. Some attempt has been made to explain the nature of Japanese corporate behaviour and its economic success emphasizing particularly static institutional factors and operational excellence. On the macro level, discussion has centred on 'Japan Inc.', the economic machine created out of the partnership between government, academic institutions, industry, and business associations. A lack of innovativeness and 'copy cat mentality' are often said to characterize the Japanese industrial circle. Likewise, much of the discussion on Japanese management has focused on bottom-up rather than top-down decision making.[4] What has been neglected is an appreciation of the vibrant entrepreneurial activity within the contemporary Japanese business world. The participation of Japanese entrepreneurs in the creation of new economic activities in the service sector in particular and the implementation and actualization of entrepreneurial aspirations deserves more attention.

'Man makes himself.'[5] The world is transformed by the decisions and actions of individuals and their institutions. One such individual, found in small and large businesses alike, is the entrepreneur – a person who is typically characterized by vision, creativity, vitality, confidence to act on new opportunities, adaptability to altered conditions and, most of all, the ability to initiate and implement change through innovation and imagination.[6] Such traits are also found within the managerial team of the innovative corporation.[7] It is this combination of characteristics that is generally accepted as comprising entrepreneurship.

The phenomenon of entrepreneurship as a distinctive human capability has attracted scholarly attention. The fact that entrepreneurs play an increasingly important role in shaping the economic and cultural landscape at corporate, national, and international levels has made entrepreneurship the frequent subject of studies in the social sciences (see Appendix, page 242).

Despite the extensive examination of entrepreneurship from various academic perspectives little has been said about the specific function of entrepreneurial vision. 'Mankind's greatest achievements are the product of vision.'[8] Vision in the business context may be defined as the dynamic mental process through which innovation is initiated and implemented through the formulation of a corporate culture which guides the business direction and activities. As Gifford Pinchot III points out,

> vision is not just a vague idea of a goal, nor is it just a clear picture of the product or service. It is a working model of all aspects of the business being created and the steps needed to make them happen.[9]

In the book *Personality in Industry: The Human Side of a Japanese Enterprise* (published under the name Hiroshi Tanaka) I identified the fact that Japanese culture and ethics are deeply rooted in cultural values and explained how influential these dimensions are in the Japanese economy.[10] Through my inquiry it became apparent that entrepreneurial vision plays a crucial role in corporate strategic design and implementation. Vision guides the organization in determination of direction in terms of products, markets, resources, and capabilities. Corporate success demands appropriate strategic thinking as much as effectiveness of operations.[11] In the present study I have identified entrepreneurial vision as the driving energy behind corporate development. I have tried to describe the reality of entrepreneurial practice as viewed and experienced by the participants themselves.

Within the rapidly changing business world of Japan, not only are many of the well-established traditional corporations adjusting to the new economic climate, many entirely new companies are emerging.[12] Moreover, many of the latter companies are now comparable in performance and size and often excel in vision and in potential, surpassing their older counterparts. Of the 1,600 corporations presently listed on the Tokyo Stock Exchange, only four have consistently shown an increase in total sales and in profits every year since 1962, the year SECOM, the company on which this book focuses, was established. SECOM is one of these four.[13] Within the Japanese business world SECOM is referred to as one of the country's 'miracle' service companies.[14]

Entrepreneurial behaviour is the antithesis of administrative behaviour.[15] Administrative behaviour is designed to optimize current operations while entrepreneurial behaviour creates change and renewal. Entrepreneurial behaviour creates profit potential while competitive behaviour exploits it.[16] In developing a corporate vision the successful entrepreneur must realize a balance between administrative functions and entrepreneurial action. Without denying the existence of administrative behaviour, Makoto Iida embodies many of the behaviourial characteristics of a successful entrepreneur. All of these characteristics are integrated into the formulation of business strategies and policies.

A 1983 study of American high-tech businesses determined that, irrespective of business size, product, and technology employed, leaders of successful high technology-based companies are confronted with the same dilemma: 'how to unleash the creativity that promotes growth and change without being fragmented by it, and how to control innovation without stifling it.'[17] The present study examines the way in which Makoto Iida has chosen to tackle this dilemma associated with corporate growth.

The book demonstrates that SECOM's progress depends largely on the corporate culture which has made possible the continuous exploration of the concept and value of safety and the corresponding development of effective and economical safety measures. SECOM's corporate culture originates in Makoto Iida and is supported by the organizational members who share his values and the corporate philosophy. As seen in this account, Iida has created, embedded, and strengthened the corporate culture through the repeated articulation of his own vision. The derivation of the company's success from the formulation and implementation of Iida's vision is demonstrated. The corporate vision provides the shared basic assumptions about the nature of the corporate mission, the product, the market, and other factors which may influence corporate development. By tracing the relationship between the ideas and behaviour of Iida and his business associates, and the realities of corporate activities, SECOM's evolution is examined.

At the time that the firm was founded the idea of peace of mind as a marketable commodity had not yet entered the Japanese collective consciousness. It would be necessary for Iida to generate a perceived need for SECOM's services. The company began with the provision of human-based, round-the-clock security. However, Iida soon recognized the limitations of this type of service and introduced a security system integrating people and technology, incorporating remote sensors at a time when such an idea was ridiculed by those engaged in security businesses world-wide. How Iida came to visualize and finally actualize the concept of machine security with limited resources is described. Expansion into machine security provided the opportunity for entry into a variety of new fields. The coordination of research and development, marketing, production, and maintenance as it relates to Iida's vision is discussed and shown to be crucial to SECOM's success in the provision of better quality service and its diversification.

From the outset, Iida acted in the international arena. Corporate establishment was made possible by an infusion of foreign currency. Although the idea of establishing a security service came from the West initially, Iida studiously rejected exposure to western security concepts and methods in a deliberate attempt to avoid imitation. Nevertheless, once a firm foundation had been established he began preparing for eventual foreign expansion. Today SECOM has expanded into seven countries including the United States, Korea, Taiwan, and Thailand. International expansion was never without difficulty and the processes through which successes were eventually achieved are outlined.

SECOM's entry into the high technology-based, on-line security system made it possible for Iida to delve into what he terms the 'social system industry'. This includes database services, VAN (value added network) service, and computer security. He has also expanded into transportation and medical services. The extent and nature of these activities are outlined.

Preparing to meet the emerging age of integrated multimedia services, today the SECOM Group consists of SECOM, thirteen domestic subsidiaries, forty-seven associated corporations, and seven foreign establishments. Total sales for 1990 approached 188 billion yen, one-ninth of the total sales of the security industry in Japan. Net profit reached 11.4 billion yen in the same year. SECOM's employees number more than 16,000. The 360,000 domestic and foreign users of SECOM's services include private households, retail stores, hotels, private and government offices, factories, schools, hospitals, banks, art galleries, museums, zoos, Buddhist temples and Shinto shrines, castles, blood banks, fish cultivation sites, industrial complexes, hydroelectric dams, airports, and nuclear plants.

It goes without saying that resources are necessary for the achievement of corporate progress, be they capital, material, people, skills, information, experience, etc. What distinguishes the entrepreneur is the ability to engage successfully in innovative ventures, even with an apparent scarcity of resources. Resources are inert. To the question, 'What is the indispensable psychic act which gives them meaning as resources?', Shackle answers, 'It is imagination, the ultimate creative act of thought in which men are tempted, with some excuse, to find their apotheosis, to see themselves as plenipotentiaries of divine power.'[18] As illustrated in this account, even mistakes and apparent setbacks can become entrepreneurial resources for growth and renewal. It is Iida's imagination and vision in the recognition of resource potential that has made SECOM's steady progress possible.

SECOM is a security firm and therefore I had to agree, at the outset of the investigation, to respect designated boundaries beyond which data collection could not proceed. Airport security, nuclear security, and computer security are three examples of areas about which I would have liked to have provided a more comprehensive portrayal but was unable to do so because of data limitations.

Another condition of the study was that the finished manuscript would be read, and its contents approved by SECOM personnel before publication. This ensured accuracy of the facts and figures from the corporate viewpoint. I was asked to delete only one section – the narration of events leading up to the extraction of the company from foreign ownership. This request came from both Erik Philip Sorensen, whose investment allowed the company to come into being initially, and Makoto Iida.

A major obstacle I encountered in my exploration was the lack of published material on the Japanese security industry. I had intended to place SECOM within the broader context of the security industry in Japan but, with the exception of the minimal data which I have incorporated into the text, the information required to do this was unobtainable. My search of the National Diet Library, a depository for all materials published in Japan, for information on corporate behaviour within the security industry, turned up next to nothing.

The alternative avenue appeared to be interviews with key personnel among SECOM's competitors. Japanese social custom prescribes that this type of meeting, like all business meetings between parties previously unknown to each other, be arranged through a third party known to both sides. I was fortunate to have made the acquaintance of a senior researcher/administrator at the National Research Institute of Police Science and with his help interviews were arranged with senior executives in other security firms. These, however, did not bear fruit. Neither did they broaden my understanding of the particular question nor disclose how SECOM is viewed in the eyes of its competitors.

Similarly, I could elicit no criticism of SECOM from the three dozen company personnel I interviewed in Japan and abroad. Had I been wearing a journalist's hat it might have been possible to enter the corporate domain through the 'back door' without corporate endorsement and to elicit from the workers their 'honest' feelings and opinions. But my focus was on Makoto Iida so I had no choice but to enter by the 'front door'. While Iida's endorsement of the project opened many doors, allowed me to see at close range his business style, and provided access to a cross-section of the company's executives, it no doubt inhibited the objectivity of the data procured.

At the time of establishment in 1962, the company was named Nihon Keibi Hosho. Twenty-one years later the name was changed to SECOM (Security Communication). Throughout the text Nihon Keibi Hosho is used in reference to corporate activities prior to the 1983 name change and SECOM is used if the discussion focuses on later developments.

Makoto Iida sought to create a business which could be beneficial to society, and through his drive and energy, attained this goal. I hope that the narration that follows will provide insight into the nature of innovative entrepreneurial activities operating in the service sector in contemporary Japan.

2 Early years

THE SETTING

Makoto Iida was born in the heart of Tokyo in 1933. The family home was near Nihonbashi, or 'Japan Bridge', in the Nihonbashi district of the city. For centuries the bridge marked the starting point of the Tokaido, the highway linking Edo (Tokyo) to Kyoto, and four other politically and economically significant highways penetrating to different parts of Honshu, Japan's main island. Road distances are still measured from this point. It was near Nihonbashi that William Adams, the first Englishman to live in Japan and a model for James Clavell's novel, *Shogun*, resided. A ship's pilot, he was called Miura Anjin (Pilot) for the twenty years (1600–20) he lived in the country. Many of Japan's oldest banks, trading houses, and department stores were founded and still remain in the Nihonbashi district. In the present century the area has become the home of the Tokyo Stock Exchange. It was in this urban commercial environment that Makoto Iida spent his childhood. The youngest in his family, he had five brothers, the eldest of whom died at the age of six months. His father was Monjiro Iida, and his mother Fusa Ohtsubo, daughter of a well-established wholesaler of traditional Japanese paper.

Ancestors of the Iida family moved from Mikawa, today Aichi prefecture, to Edo, after Ieyasu Tokugawa (1542–1616), who was to become the first Tokugawa Shogun, was forced to move from Mikawa to Edo. When this occurred Ieyasu brought with him, in addition to his vassals and their families, a number of purveyors including manufacturers and traders of guns, pottery, cake, and sake, and also carpenters. Ieyasu's new domain effectively embraced the Kanto Plain and its surrounding hills. It was geographically more unified, larger, and considerably more productive than the old domain. The next fifteen years were spent preparing Edo for its future role as the nation's 'capital'. Ieyasu ordered the construction of a castle, a network of canals, a water supply system, and land reclamation through the drainage of the swamp which lay between the town and the sea.

To this new site Ieyasu welcomed commoners from other parts of Japan, especially from his own home provinces of Mikawa and Suruga, presently Shizuoka prefecture. The area east of the castle was assigned to the merchant class as its numerous canals allowed for the export and import of commodities to and from other parts of Japan. The Shogunate government bestowed protective privileges on the merchant class which was concentrated in the Nihonbashi and Kyobashi districts of the newly reclaimed area of Edo. Members of the merchant class for several generations, the Iida family was financially established. Indeed, Iida's ancestors had built an extensive trading business in oils to be used in a variety of different types of lantern.

For 250 years following the establishment of the seat of government at Edo the country was closed to foreigners. The Tokugawa reign was a time of prosperity and peace. A monetary economy emerged. Social development resulted in increased oil consumption, especially for lighting. Initially only upper class samurai and wealthy merchants could afford the luxury of lighting their homes at night but gradually, as the standard of living rose, commoners were able to enjoy this convenience as well. Places of entertainment also adopted oil lighting. In order to meet the rising demand, the Tokugawa government encouraged the cultivation of rapeseed for oil.[1] Among the enterprising merchants in Togukawa Japan, Iida's ancestors were involved in the distribution of rapeseed oil from its source in central Honshu to buyers in the nation's capital.

Towards the end of the Edo period (1603–1868) the Tokugawa government opened select ports to foreign ships. With this change the bright lights of the West came into Japan. In Yokohama and other port cities the residences and offices of western merchants were brightly lit by oil lamps. In the early 1870s gas lights were introduced on the streets of Yokohama and Tokyo and towards the end of the decade their use had spread to private homes. By the end of the 1880s gas and electric lights were common in these centres and the use of oil for lighting was in decline.

The Meiji Restoration of 1868 and the subsequent exposure of Japan to foreign ideas and technology brought dramatic changes in almost every aspect of Japanese life. Oil merchants, like everyone else, had to adjust or disappear from the scene. Under these conditions Iida's great grandfather, Shinsuke, ceased dealing in oil. Sake, soya sauce, and miso (bean paste), all of which shared the same distribution channel, became the commodities of the small retail business. Iida's great grandfather chose sake as the primary commodity of trade because of the great demand for it. When he made the transition from oil to sake about 700,000 barrels of the liquor were being shipped to Tokyo annually from production points across the country.[2] This was at a time when the appropriate social role of the entrepreneur was the subject of fervent discussion.

The Meiji Restoration marked the beginning of the legitimization of entrepreneurship in Japan.[3] In order to maintain the feudal social structure the Tokugawa Shogunate had relegated the merchant class to the lowest rung in the hierarchy, after samurai, farmers, and craftsmen, and their position improved only slightly throughout the three hundred years during which the era continued. The low esteem in which merchants were held was supported by the Confucian disapproval of economic activity which placed individual before societal benefit. The Confucian emphasis on devotion to duty, however, may have supported entrepreneurship in Japan just as the worldly asceticism of the Protestant Ethic supported European entrepreneurship. While the latter was for the direct benefit of the individual and only indirectly of society as a whole, the former was directly oriented to the benefit of the group.[4]

As the Meiji era progressed the climate for entrepreneurial activity became more favourable with strengthening government support in the form of the construction of model factories, the privatization of its enterprises, subsidies to entrepreneurs, and the promotion of technological innovation. Entrepreneurs themselves sought to gain prestige through a two-pronged campaign in which 'they presented themselves as the embodiment of samurai ideals, devoted to the tenets of Confucianism.'[5] They were working for the betterment of society. Private profit, they insisted, was of no interest. It was at this time that Eikichi Shibusawa, a leading figure in the campaign, coined the term *jitsugyoka*, or 'a person who works with honesty for the establishment of industry', for the entrepreneurial role.[6]

In a social and economic climate that favoured the entrepreneur, Iida's ancestors' business continued to thrive and grow and was passed down first to Iida's grandfather and then to his father. When Eikichi Iida, the grandfather, assumed control of the business he named it Okamotoya, a name Iida's father later changed to Okanaga. With the name change came a corresponding change in the nature of the business from a retail to a wholesale operation. Handed down as well was the entrepreneurial belief that business was first and foremost for the benefit of society. Iida has few memories of his grandfather who

1 The Iida family, 1942. Makoto Iida (far left) with his parents (seated) and his four brothers.

died when Iida was six but he remembers his father as an innovative, decisive, and upright businessman, meticulously honest in all his business dealings.

In its advantageous geographical location the business flourished under Iida's father's leadership. By the late thirties, when Iida was a child, the business had expanded to employ fifty workers who lived in close proximity to the family. Five maids performed the household chores for the workers and the Iida family. In addition, the family owned real estate and employed several people to collect the rent and manage the properties.

Iida and his brothers, Hiroshi, Tamotsu, Susumu, and Atsushi, were close in age with only ten years separating Iida, the youngest, from Hiroshi, the eldest. They were rivals in their academic as well as recreational pursuits. Their home, Iida remembers, often resembled a battlefield with toy swords strewn everywhere. Within this environment Iida learned to be assertive in order to hold his place in the competitive arena. The active family setting in which he grew up helped shape his cooperative, but at the same time independent, character. Says Iida's eldest brother, Hiroshi: 'Makoto was the youngest and we all loved his friendly nature. His outgoing personality brightened up the family.'

Business endeavours prevented Iida's father from spending a lot of time with his sons but occasionally he would take his children on an outing. Being the youngest Iida would often be the only one to accompany his father. Iida remembers that on such occasions when he would tire after the day's adventures and, as other people around him, assume the common outdoor Japanese resting posture – a squat – while waiting for the train home, his father would reprimand him. For the senior Iida, to squat was to admit defeat. The physical action of squatting, he felt, not only indicated submission to adverse conditions but also dislocated one's spiritual stance. He taught his sons that regardless of how tired they were, their choice of posture was always a conscious decision. He insisted that they remain standing until they could sit 'properly'. As Dubos says,

> Human freedom does not imply anarchy and complete permis-siveness. . . .Design, rather than anarchy, characterizes life. In human life, design implies the acceptance and even the deliberate choice of certain constraints which are deterministic to the extent that they incorporate the influences of the past and of the environment. But design is also the expression of free will because it always involves value judgements and anticipates the future.[7]

Iida's father's insistence that he refrain from squatting had a profound effect on Iida's future behaviour. It taught him that one need not be directly influenced by conditions. There is always a choice: one can control one's own destiny or one can allow outside forces to control it. By making deliberate, conscious decisions one can 'win', in the sense of taking control of the situation. It is design and vision that gives one that control. Says Iida:

> There are times when I would like to let the business rest, maintaining and defending the present situation only, instead of relentlessly pushing forward but the lesson I learned from my father – never to squat – is a symbolic reminder never to give in to circumstances. Vision plays an important role in guiding the corporate advance.

SCHOOL LIFE

Iida was in the sixth grade, the final year of elementary school, in 1945. World War II was taking its toll on Japanese civilians as well as its military forces. Submarines of the United States Pacific Fleet had ravaged Japanese shipping fleets. Although the Dutch East Indies were still largely under Japanese control transportation was insufficient to allow anything more than a trickle of oil to enter Japan. The supply of rice from Indochina, Japan's main source of this dietary staple, was similarly cut off. As a result food was in short supply throughout the country. The mainstay of Tokyo residents was the sweet potato.

As American victories in the Pacific secured air bases for the Allied Forces within attacking distance of Japan the likelihood of air raids on Tokyo escalated. School children were collectively dispersed to nearby rural areas. Iida's class was sent to Nakuri village in Saitama prefecture north of Tokyo where accommodation was provided in temples, schools, and with the villagers.

A straight 'A' student who excelled in mathematics, Iida had made application to take the entrance exams for a middle school in Tokyo. However, on 25 February 1945 downtown Tokyo was extensively bombed. B-29s dispatched from the American air base in the Marianas dropped over 400 tons of incendiary bombs on one square mile of urban Tokyo.[8] The area in which the Iidas' home and business were situated was razed in the resulting fire storm. This was only the beginning of an intense series of air raids that would level much of downtown Tokyo and other major cities. With the home and business gone the Iida family moved to their summer home in the seaside resort town of Hayama in the Shonan district of Kanagawa prefecture south of Tokyo and Iida was reunited with his family. Despite the nationwide food shortage the family ate well, Iida remembers, on the results of the children's daily marine harvests. Among other seafood, sole and gurnard, today expensive delicacies, were dietary staples.

Iida subsequently sat the entrance exams for Shonan Middle School, later, with the American postwar reform of the education system, to become Shonan Senior High School. In the interview that followed the written examination Iida was asked about his career intentions. At that time he had his heart set on becoming a naval officer. The Yokosuka Naval Base was 10 kilometres east of Hayama and several officers had residences in the area. Many graduates of Shonan Middle School attended the Naval Academy and the current school body included the sons of a number of naval officers well-known for their war exploits. The naval influence was strong at Shonan Middle School.

Iida's proud announcement of his intention to become a naval officer elicited a sigh of despair from a member of the examining committee. A military career was the dream of all the young boys. Was no one interested in becoming a diplomat or *jitsugyoka* – an entrepreneur – the disgruntled interviewer had wondered aloud. Hearing this, Iida feared he would fail the examination but, instead, he was accepted into the school. Nevertheless the interviewer's statement made a lasting impression on him.

On 15 August, four months after Iida began middle school, Japan surrendered unconditionally to the Allied Forces. The capitulation came as a tremendous shock to the 12 year old. He had been taught that Japan, the land of the gods, was invincible. Now the myth had been shattered. The indoctrination received during the war years left him in terror that he would now be killed by the Americans. But this did not happen. Suddenly people were talking about various political ideologies – democracy, socialism, capitalism. It was impossible for him to grasp the direction in which the country was moving in this ideological chaos. It was out of Iida's experience of the war that his inclination to question established values developed. This questioning attitude would nurture innovative thinking. 'There is nothing on this earth that has to be the way it is. Established customs can be changed. This belief is rooted in my childhood experience of the war.'

While struggling to reformulate his value system, a severe bout of enteritis necessitated Iida's repetition of grade seven. For three months he commuted twice weekly to a gastroenteric hospital for treatment and he was finally cured by moxa cautery. In Iida's class the second year was Kei Kimura who would later become president of SECOM.

Soon after the end of the war Iida's father and eldest brother returned to Tokyo. They erected a shack on the family property but were prevented from resuming business for some time as sake was a controlled item in the postwar economy. With the rest of the family Iida remained in Shonan district and after their return to Tokyo commuted daily to Fujisawa to complete his high school education.

Bewildered by the feelings of loss elicited by the defeat and wrestling with the rapidly changing social values, Iida found a vent for his emotions and energies in Shonan's natural environment – the Pacific Ocean, the sun, the mild climate, and the beautiful setting with Mount Fuji in the distance. He took up sailing and developed an intense love of the ocean. Throughout his life this love has remained with him. Even today, when he is overcome with stress he travels to the seaside and lets the expansiveness of the ocean wash over him. Soaking up the fresh air either on shore or on his fishing boat, within an hour he will feel renewed 'The ocean provides a setting in which I can think without interruption. There I am in an environment completely different from that of my everyday life.' To Iida the ocean is a sanctuary to which he can retreat for the re-creation and revitalization of mind and body. It is a place of pilgrimage where, 'I become a tiny speck in the expansive seascape.'

Iida's circle of friends expanded rapidly and included, among others, Shintaro Ishihara who was later to become a novelist and in 1972, a member of parliament. A student at Iida's high school, Ishihara went on to enrol in the Faculty of Law at Hitotsubashi University in Tokyo. While there he wrote the novel *Taiyo no Kisetsu* (Season of the Sun) about the carefree life of youth in Shonan district. The novel won the Japanese Authors' Association New Author Award in 1955 and the prestigious Akutagawa Prize in 1956 and brought the term *Taiyozoku* or 'Sun Tribe' into common usage.

Taiyozoku symbolized the emergence of a new age in Japan. Standing in sharp contrast to the drab, harsh struggle to survive experienced by most Japanese in the years immediately following the war, *Taiyozoku* were associated with youth, flamboyancy, vitality and the negation of established values. While Iida denies that the novel is an accurate reflection of his personal situation at the time, he certainly experienced the free, easy-going, optimistic life described therein. Even today, from time to time, the Japanese media will refer to him as the *Taiyozoku* who made it in the business world.

Shonan High School was well known for its academic excellence and for its performance in team sport, particularly soccer and baseball. Iida enjoyed baseball but only the very best players were given much opportunity on the field. In order to have more immediate involvement with a sport, Iida founded the school's rugby team. This was perhaps the first indication of his ability to recognize an opportunity to start a new 'venture' and to follow through successfully. To compensate for the shortage of team members, for matches, players were recruited from the soccer team. Widespread shortages of all commodities meant that the new team was without uniforms so they played in their old clothes.

Four decades later Iida would establish a company rugby team and build a state-of-the-art playing field for his employees. As coach he has recruited Shinichi Hashimoto of Russian/Japanese parentage, a once famous player with the Waseda University rugby team. Blessed with many members who are former high school and university rugby players and with excellent facilities, the team does very well in corporate competitions.

Iida has always enjoyed the challenge of starting new activities be they in the sports or business context. At Gakushuin University,[9] where he was enrolled in the Faculty of Political Science and Economics, he first joined the rugby team and later, in conjunction with two senior students, started an American football team.[10]

When Iida founded Gakushuin's football team the university had little to offer in the way of human or financial resources. It was the smallest of all the universities to have a team. The first challenge with which Iida was confronted was recruitment. There were barely enough players to have a game. As the captain Iida worked hard to entice freshmen to join the team. His recruitment strategy included frequenting off-campus tearooms, a popular gathering place for Gakushuin students. Hisaki Shimosato, two years Iida's junior and now an executive director of SECOM, says of the situation:

> I used to go to a Shinjuku tearoom with my friends. There was always a crowd of Gakushuin students there. Among them was a tall, lively individual. This was my first encounter with Iida. Students often spoke of his likeable personality and cheerful disposition. Students, especially the younger ones, tended to follow him. Iida's parents gave him spending money in addition to what he earned at a variety of summer jobs and won at the *pachinko* parlor.[11] If anyone ran short of funds he'd loan them money for rent and/or tuition. He would often buy us freshmen meals and coffee and talk to us about the importance of being fit and staying healthy. I had played touch football in high school and, attracted by Iida's personality, signed up for the university team. There were thirteen of us from my freshman year who joined the team, bringing the total number of members to twenty-four. We were then able to be officially registered as the university football team; however this new status gave us neither a coach nor sufficient money for equipment.

Having recruited the necessary players, Iida's next challenge was to procure the funds necessary to purchase uniforms and equipment. Lacking even blocking equipment for tackle practice, initially Iida served as a human dummy, insisting that his teammates tackle him. Those who restrained their tackle for fear of injuring Iida would incur his wrath and have to serve as dummies themselves. Iida secured some funding from the university to augment that raised through team-sponsored dances but it was still not nearly enough to purchase uniforms and the necessary equipment. However, this problem solved itself.

An American businessman, in Japan initially with the military, observing the team's practice one day, inquired as to whether or not they had a coach. When the response was negative he offered his services. Later he arranged for the donation of used footballs,

uniforms, and equipment from the football teams of the American
military bases in Japan. Iida borrowed his father's truck and travelled
to each of the bases to collect the donations. The Allied Occupying
Forces, under the programme designed to put Japan on the road to
recovery after its devastating defeat, were happy to supply the
students. 'American sports' were perceived as one avenue through
which the seeds of American culture could be planted in Japan.
Military personnel also made available to Iida books on football
strategy. Iida thought that his team was the best equipped and best
dressed of any varsity football team in the country. Before this could
be, however, the uniforms and equipment had to undergo considerable
alteration to scale them down to size for the Japanese players.

As team captain Iida felt responsible for his teammates' academic
performance. He insisted on their attendance at lectures as well as at
football practices. Older players would advise the younger ones on
course selection and lecture notes were made available. Nevertheless,
there were still a few whose academic performance was unsatisfactory.
Iida would accompany such an individual to the appropriate professor's office where he would kneel Japanese fashion in front of the
teacher while he and Iida reprimanded the offending student and
commanded him to study harder. Most of them graduated eventually.
When anyone broke the rules Iida would always accompany the
offender to make amends. Says Shimosato, 'He placed importance on
proper etiquette in dealing with faculty and senior students and
showed us how to behave.'

Correct behaviour towards all those with whom one is in contact is
essential in Japan. As Nakane explains:

> A Japanese finds his world clearly divided into three categories,
> *sempai* (seniors), *kohai* (juniors) and *doryo*. *Doryo*, meaning 'one's
> colleagues', refers only to those with the same rank, not to all who
> do the same type of work in the same office or on the same shop
> floor; even among *doryo*, differences in year of entry or graduation
> from school or college contribute to a sense of *sempai* or *kohai*.
> These three categories would be subsumed under the single term
> 'colleagues' in other societies.[12]

The type of relationship that exists between two individuals influences
every aspect of their interaction, particularly their verbal and non-
verbal communication. Social education begins in childhood and
continues into adulthood. Correct social behaviour is expected and is
essential throughout life. The Japanese student sports team provides
an appropriate environment in which these skills can be honed.

Growing up in a business family and having four older brothers Iida was acutely aware of the importance of correct social behaviour and felt it essential to impart this knowledge to his teammates. Says Shimosato, 'Iida was respected not only by his team members but by the university faculty and administration as well.' His open mindedness, willingness to learn from others, and highly accomplished social graces equipped him well to interact with his seniors in the business world in later years. Others would seek him out to share their knowledge with him and to learn from him.

Iida's business success is attributable in part to his ability to communicate with and motivate a wide assortment of people of different ages, educational, professional and cultural backgrounds, and personality. During their long association, Michael Kaye, who heads SECOM's American operation observed that, like any successful leader, Iida is different things to different people, wearing different faces, each one appropriate for the individual to whom he is relating.

> It is necessary to relate differently to different people. He wears many different faces, yet they are all Iida. He can play those different roles which he needs to do to motivate a lot of different kinds of people. But at the same time there is a universal quality about him.

Despite Iida's efforts and those of the football team's American coach, the team did not perform well in matches. Nevertheless, exposure to this sports endeavour provided an excellent learning environment for all those involved. After each game strategies and plays would be analysed to provide the basis for planning the next game. The team used eighty standard plays, considerably fewer than other universities, but with less than a full complement of players Iida had to give careful thought to the best use of people in each formation. For Iida, it was the planning/action/evaluation aspect of football together with the game's constantly changing dynamics that fascinated him. The systematic, contingency approach that football strategy demands was in accord with Iida's personality. The knowledge and experience gained in organizing the team and participating in the sport would benefit him greatly in his later business life.

FAMILY BUSINESS

During World War II sake was a rationed commodity with production and distribution tightly controlled by the Japanese government. After the war the American government contemplated a temporary ban on sake production and sales. When the Japanese pointed out the need for the taxes from sake sales to replenish the empty government coffers the United States relented and authorized controlled production and distribution. With the extreme shortage of liquor, methyl alcohol, salvaged from bombed factories, became available on the black market. Its consumption resulted in numerous cases of blindness and some deaths.[13]

While other sake brokers made substantial profits handling rationed commodities on the black market in the years immediately following the war, Iida's father refused to be involved in illegal activities. For this reason Okanaga did not deal in sake again until the economic controls were lifted in 1949. Like most Japanese immediately after the war, the senior Iida was willing to involve himself in whatever type of legal business he could in order to survive. His establishment resembled a furniture store and sold whatever used and new furniture and housewares were available.

Government regulations and restrictions may be viewed as having a negative effect on the entrepreneurial desire to work and to innovate. But the opposite may also be true: government imposed limitations for the public good may stimulate entrepreneurial activity through competition for innovations that satisfy social needs while conforming to the limitations. As Kurz points out, 'regulations and legal limitations on free business activities may also induce a very active entrepreneurial activity in the black or illegal markets.' [14] Many wholesalers and other merchants were engaged in black market activities as were *boryokudan*, members of organized gangs. The latter used the black market to accumulate wealth and gain strength in order to expand their territory.[15]

In postwar Japan, under a tightly controlled economy, those entrepreneurs licensed to conduct business needed to be resourceful and innovative in order to utilize effectively the limited resources available. Among racketeers who chose to enter the black market the entrepreneurial spirit ran high, fed by the need for ingenuity in both the production and distribution of controlled goods.

Iida recalls visiting his father shortly after the war. The senior Iida, clad in a tattered jacket, was cooking a meal over a charcoal burner outside the shack that was his Tokyo home. Iida was struck by the contrast between his father and other apparently more successful wholesalers. For Iida's father, the entrepreneurial spirit had to find expression in sanctioned societal activities. As if he could read his son's mind, Iida senior pointed out that profits could be easily had if one chose to be involved in the black market but that success based on illegal activities would be short-lived: honesty and integrity are indispensable virtues of the entrepreneur. As a descendant of an Edo merchant family, the necessity of 'proper conduct' in commercial affairs was ingrained in his soul.

> I can still see my father standing by the fire. What he said that evening had a profound affect on me. It was then that I knew that I would never engage in entrepreneurial activity for the sole purpose of making a profit or even making a living. My future business endeavours would have to make a societal contribution first.

While he was waiting to re-establish his sake business, Iida's father, a proponent of new ideas, encouraged his three eldest sons to try out other business ventures. The two youngest children were still students. The eldest son, Hiroshi, an economics graduate of Tokyo University, became involved in the sale of bamboo construction materials. Tamotsu, the second son, a graduate of the Department of Agricultural Economics of Hokkaido University, purchased a railway carload of prunes and a large cooking pot. Sweets were in short supply in Japan and the prunes, cooked with a quantity of sugar, were an immediate success. He later expanded into cake and then chewing gum production. Capitalizing on the popularity of the elephant in Ueno Zoo in Tokyo, a gift from Prime Minister Jawaharlal Nehru of India, Tamotsu successfully marketed *zo* or 'elephant' gum. Susumu, the third son, left the navy after the war and worked as a lacquerware wholesaler.

When the economic control on sake was lifted the senior Iida, in conjunction with his three eldest sons, re-established the family wholesale sake business. He was one of the nation's 4,500 sake wholesalers to be granted a licence in 1950. The 4,500 wholesalers represented only about one-third of those of the prewar period.[16]

Sake was still a scarce commodity and unscrupulous dealers would dilute the product in an effort to realize larger profits. This gave rise to the term *kingyoshu* or 'gold fish sake', which was so weak that gold fish were reputed to be able to swim in it unaffected. Together with the gradual economic recovery and the availability of rice came increased sake production and consumption and the emergence of a myriad of retail and wholesale sake dealers. The excessive competition resulted in price wars and credit sales, bankrupting many dealers. Iida's father skilfully steered the business through these difficulties and it flourished. By the time Iida graduated from university in 1956 the employees numbered sixty.

Iida was unsure of what should be his future. Most of his friends were preparing for jobs with large, established corporations. His father wanted him to join the family business but he had an overriding desire to be independent. He could not forget the urging of his middle school examiner many years earlier to become an entrepreneur. He had no desire to work for anyone but himself. His thoughts of being his own boss focused on the feasibility of opening a chain of ice cream stores. Ice cream was a relatively new commodity in Japan. His father was against this idea and insisted that he join the family business, a decision that baffled Iida's brothers. Three were already working for their father and they felt that the size of the wholesale operation did not warrant the addition of another family member.

At the time, Iida could not fathom his father's motives. With the passage of years, however, he has come to believe that his father, who was 65 years old when his son finished university, wanted to pass on to his youngest child the considerable business know-how he had gained from his own years of experience. The senior Iida knew that he was getting old and that the time available to him to be actively involved in his son's business education was limited. Rather than send his son to another corporation to learn the skills necessary eventually to run a business, the father chose to take full responsibility for this preparatory education. Iida's brothers had married and were living away from the family home by the time Iida joined Okanaga. The time had come for him to give up his 'carefree' student life and, as the newest employee, submit himself to the strict supervision of his father.

3 Business experience

LIFE AS AN EMPLOYEE

About ten new university graduates started to work for Okanaga concurrently with Iida. They were assigned first to the warehouse, unloading and loading sake barrels and bottles as shipments were received and orders dispatched. Gradually the other young men were given different jobs and Iida alone remained as a warehouse worker for two years. Iida was of an easy-going temperament but after a time he began to question his future. Would his entire life be spent hoisting sake bottles?

It is common practice in Japan for future managerial candidates, as a part of new employee education, to undergo on-the-job training that gives them practical experience in various facets of corporate activity. For anywhere from a few days to several months university graduates may be involved in labour intensive activities working alongside regular employees often in round-the-clock shifts. Involving these new employees in this exercise is considered to enhance their development as future managers.[1] In the case of Okanaga, loading and unloading sake was central to the company's operation. Nevertheless such manual work could be left to other labourers. Iida did not object to being assigned to the warehouse for experience but two years was unduly long, he felt.

While Iida found the work strenuous and the day long, it was not as difficult as the period that followed over a late dinner and afterwards with his father. As had been his custom for many years, his father would change into kimono and relax beside the large *hibachi*, eat *sashimi* (prepared raw fish), drink sake and talk about a variety of things. Iida had to listen attentively to his father's countless repetitions of stories of his business life – how to conduct business, how to use money wisely, and how to live one's life as a human being. The lecture would continue until his father eventually fell asleep on the floor. Iida found the ordeal, night after night for the first three years after he joined the company, psychological torture.

Hard as they were to tolerate at the time, Iida feels that his father's lectures did contribute to the conceptualization of a fundamental business paradigm. This would become the benchmark against which his future business decisions would be made. The concepts Iida gleaned from his father's teachings were fundamental. In order to operate a business beneficial to society, the entrepreneur's approach must be bold and direct and in accordance with prevailing business ethics. To keep abreast of constant change and propel the company forward, one must reject continually the present situation and seek a new direction. Corporate vision will be revealed only through concentrated conscious effort.

Iida's father did not talk exclusively about business. Sometimes he would hold forth on how to interact with people, sometimes on the proper appreciation of sake. Sake drinking has been enjoyed since ancient times. The custom was initiated in the court and handed down to the samurai class and eventually to commoners. It is a ritual that often accompanies occasions such as festivals, gatherings, weddings, funerals, and coming of age ceremonies. Its function is to relieve tension, to signify approval, to indicate appreciation and to express care and concern, to strengthen bonds of trust, to heighten team spirit, to improve communication, and to signify the end of an association between individuals or between individuals and an organization. The manner of offering, accepting, and enjoying sake varies with the occasion and the people with whom it is shared.

Being in the sake wholesale business, the drink was abundantly available. Had they wished, the senior Iida and his sons could have continually indulged themselves. He urged them, however, to enjoy sake but never to misuse it – to enjoy the time, the place, the company, and the sake but never to use sake as a release from stress and as a 'solution' to problems. The advice was common sense and it has stayed with Iida through the years. Whenever the opportunity arises he passes it on to his employees. Iida's early familiarization with 'sake rituals' would prove useful not only in Japan but in the conduct of business in East Asia.

Although the lessons Iida learned from his mother were more subtly proffered, they too were to play an influential role in developing his attitudes. Throughout his life, from early childhood on, his mother has reminded him of the need for humility regardless of his business achievements. While recognizing that self-confidence is essential to entrepreneurial success, she is aware that it often breeds arrogance and counsels him to avoid the latter, to remain humble always. As a child, Iida worried a great deal. From time to time he would sigh audibly from the burden of his cares. Gradually his mother taught him the value of a positive outlook. He learned that throughout life there would always be things to worry about. But by viewing them positively he found that he could meet them head on and deal with them successfully.

Two years after he joined the family business he was transferred from the warehouse to the sales department. By this time his eldest brother, Hiroshi, had been made vice-president, Tamotsu was senior executive director responsible for sales, and Susumu was executive director. Despite the move, the pattern of his life changed little. Before dawn Iida's father would often wake him with shouts of 'Don't you understand my teaching?' He would remind Iida that he should be up and preparing to go and collect the debts owing to the company; that at this early hour he would be sure to catch the household/business heads at home. Iida would rise soon after, silently protesting that it was still too early, and prepare for the day's work.

His job at that time was concerned with selling sake and food commodities to department stores, large grocery stores, and smaller wholesalers. Iida's department, especially the section concerned with small wholesalers, was plagued by the collection of monies owing. Because of the intense competition among them, small wholesalers throughout the country were in a tenuous position, living almost from hand to mouth. Bills were paid only to the most demanding creditors. By insisting that Iida stay on top of debt collection, his father taught him one very important aspect of business management. It also instilled in him an intense dislike of credit in any form. In his later business dealings he would insist on payment-in-full when the contract for service was drawn up.

While most business difficulties were eventually worked out between the parties involved, it was sometimes necessary to go to court in order to settle differences. On such occasions Iida would often be accompanied by Yasuharu Nagashima, a young lawyer who was a close friend of his brother, Tamotsu. The relationship between Iida and Nagashima has continued to the present day with Nagashima acting as a lawyer for SECOM on many occasions.

Although Okanaga was the focus of Iida's existence his life was not all work. On Sundays he would often take in a musical or a play with a number of friends including Shinichiro Osada, a friend from high school days who would later become a senior managing director of SECOM. Like Iida, Osada had lived in Tokyo until the wartime evacuation when he had moved with his family to the outskirts of Fujisawa, the town to which Iida travelled daily from Hayama to attend school. Osada remembers Iida as a well-mannered, quiet boy but yet mischievous, above average in his academic pursuits. Iida and Osada became close friends. It was Iida who was always asked to take messages from the school to Osada's residence when his friend was absent from class – a frequent occurrence as Osada had a habit of skipping school on Mondays and rainy days.

When Iida's family returned to Tokyo after the war Iida commuted to Fujisawa to finish his schooling. Later, although Iida and Osada attended different universities in Tokyo, they would continue to meet. Osada would often visit Iida at his part-time job in Okanaga's warehouse. As was Iida's custom when his friends came to visit, the two young men would drink sake together, claiming later that the barrel had leaked.

After graduation Osada joined Nippon Television Broadcasting Corporation. He continued to visit Iida who was now fully employed by Okanaga. When he would reminisce about their consumption of sake in their university days and suggest that the tradition be maintained he would meet with strong opposition. Iida had listened many hours to his father's lectures on business management and his attitude to the company's merchandise had changed dramatically. Osada remembers that when the weather was cold Iida would often stop by the television station, if it was on his debt collection route, and the two would have coffee together. Iida would discuss the difficulties of collecting monies owing. Osada recalls, 'Iida was always very well dressed except for his worn and scruffy shoes. You could tell from his shoes that his work necessitated considerable walking.'

At about the same time that Iida was transferred from the ware-house to the sales department of Okanaga he met his future wife, Misao Kakuta. They were to be married, one year later, in 1959. Shortly before the wedding Iida realized that he did not have a pair of black shoes for the ceremony. 'I didn't have the money for a new pair of shoes and I didn't want to ask my father for a handout so I had to improvise. I put a thick coat of black shoe polish on my brown shoes and managed to complete the wedding rituals.' In 1960, Iida's first child, a son, was born. Iida's father named him Goichi, meaning 'the strongest one'. Iida and his wife and son had their own residential quarters on his father's property, relieving him at last from regular attendance at his father's after dinner lectures.

Iida had wished for independence in business from the outset. Marriage with its attendant responsibilities did not deter him from this goal. After some time in the family business he started to imagine himself at age 50, still living on the family property and working for his brothers. He felt like a corralled race horse champing at the bit to run. The desire for independence was shared by Iida's second and third brothers, Tamotsu and Susumu.

Their father kept the sons close to him while at the same time urging them to become self-employed. The first to strike out on his own was Susumu. After the war, Susumu spent three months in the United States where he observed the rapid diffusion of the supermarket. Upon his return to Japan, Susumu established a new section in the company to deal with a variety of items other than sake. He intended eventually to break away from Okanaga and establish this former section as an independent business. In the mid-fifties there was considerable criticism of wholesalers who became discount supermarket retailers. His father, fond of innovation, encouraged him to go ahead with his plan and in 1954 Susumu opened his first supermarket. Now, in the early nineties, Susumu heads up a large, successful, forty-member chain of supermarkets centred on the outskirts of Tokyo.

Susumu's newly acquired independence inspired Iida and his brother Tamotsu to follow suit. It was assumed that Hiroshi, the eldest brother, would run, and eventually inherit, the father's business. The knowledge Iida acquired working for his father for several years combined with his taste for the excitement of creating new organizations gleaned from his experience with the establishment of sports teams in his youth heightened Iida's drive toward independence. He began to give serious thought to what type of business would be most appropriate for him.

FRIEND AND PARTNER

Iida's search for an appropriate channel for his energies encompassed both the domestic and foreign scene. He had one friend, in particular, with whom he shared his search. Juichi Toda, later to become executive vice-chairman and representative director of SECOM, was a year older than Iida.

Toda's father, Toshiharu Toda, had emigrated to Canada from Yamanashi prefecture in central Japan in 1907 at the age of eighteen, one of 2,753 Japanese to do so that year.[2] He subsequently took Canadian citizenship. He first attended technical school at night in Vancouver and later operated a fishing business and a small hotel in the city. He enjoyed writing *haiku*, seventeen-syllable Japanese poetry, and formed a *haiku* association. Seeking a spiritual anchor in the new land he explored various religions and eventually converted from Buddhism to Catholicism.

Not finding a suitable bride among the few single Japanese women in Canada, he selected a wife from photographs of Japanese women. Tsugi Kai came to Canada from Kagoshima prefecture and they were married shortly after her arrival. Toda's eldest sister was born in 1922. In 1929, seeking a healthful climate in which to recuperate after a serious illness, Toda's father moved his family inland to the town of Okanagan Centre on the eastern shore of Okanagan Lake in the interior of British Columbia.

It was here that Toda was born in 1932, the fourth and last child and only son. Baptized a Catholic he was christened Juichi Joseph Toda. When his father died four years later, his mother, who had maintained her Japanese citizenship, moved the family back to Japan. Toda clearly remembers his father on his death bed saying to him, 'You are the only male in the family. Please look after your mother.'

Returning to Japan the family lived near Toda's father's brother, Tatewaki Laurence Toda, a Catholic priest in Tokyo. He had converted to Catholicism after consultation with Toda's father, and subsequently joined the priesthood. A pacifist, the uncle protested against Japan's war effort and was jailed several times for his objections. He was particularly vocal against the militarization of Yamate Church in Yokohama. Taking advantage of the church's strategic coastal location, the Navy installed cannons in its precincts as preparations were made for the landing of American forces on Japanese soil near the end of the war. In August 1945, three days after Japan's surrender, Japanese military police burst into Hodogaya Church and shot and killed Laurence Toda while he was praying.[3]

Toda was 13 years old. He remembers advising his mother against trying to bring the murderer to justice, a futile task in the turmoil that followed on the heels of Japan's defeat. Toda bore no hatred towards the murderer, nor did he have any strong emotions regarding his uncle's death. The overriding feeling was one of emptiness.

My father had died and the uncle who had taken his place was dead also. There was no longer anyone for me to rely on. This fact together with Japan's defeat and the subsequent disintegration of the nation's 'adult society' made it necessary to develop self-sufficiency and independence.

My mother followed my father and uncle in their acceptance of Catholic beliefs and prayed frequently to God that I would grow up to be a respectable person and not a burden to anyone. My attitude was to accept life as it came and I rejected my mother's overriding dependency on God. I wanted to distance myself from her and was acutely embarrassed by her occasional visits to my school. Thinking back, I was an uncaring child.

From my father and mother I learned about the importance of family life and from my uncle I acquired the ethical stance by which I live.

After completing his elementary and high school education in Tokyo Toda enrolled in the Faculty of Political Science and Economics at Gakushuin University a year prior to Iida. With Toda being on the baseball team and Iida on the football team, their paths often crossed and they became good friends. After completing his studies at Gakushuin, Toda had hoped to pursue graduate studies abroad, however, when his mother became ill he found it necessary to remain in Japan. Having planned to go abroad Toda was not among the new graduates seeking employment to commence 1 April, the one day in the year when most new employees begin work. Good jobs were scarce in the mid-fifties and, having missed the time for formal entry into the work force in his graduating year, the best Toda could do was to secure a low paying part-time position with a travel agency issuing tickets and processing travel documents.

When he ran short of money he would go to Okanaga and help Iida with his work in the warehouse. When the work was done the two men would eat and drink together and talk about Iida's future independence, giving careful thought to what might be an appropriate business venture for him. As he had done from the time Iida and Toda were in university, Iida's father would often join the two young men and 'lecture' them about the conduct of their lives. Because Toda had been without a father most of his life, he enjoyed being included and developed a great fondness for Iida's father. The feeling was mutual. The first of the senior Iida's sons who had died at six months of age had been named Juichi. Iida's parents found a special place in their hearts for Juichi Toda.

Over several months the two young men came to the realization that they wanted to go into business together. From the outset it was understood that Toda would support Iida totally in the operation of the business venture. Toda has always done so, placing Iida in the limelight and taking on whatever roles Iida wishes him to assume. Says Iida, 'My mother still reminds me from time to time how fortunate I am to have Toda as a friend and partner.' Throughout the course of the company's development Toda has played two important roles. He assists Iida in implementing ideas and plans, following through and finishing what Iida starts, thereby freeing his friend to go on to other things. In addition, Toda acts as a sounding board for Iida's new ideas viewing them from a different, sometimes more practical perspective. His opinions sometimes differ from Iida's. Iida takes Toda's advice very seriously and re-examines his plan. The two men each recognize the unique strength and role of the other and respect each other wholeheartedly.

DECIDING ON A VENTURE

Whatever entrepreneurial undertaking Iida might settle on, there were three conditions he insisted it meet. Above all, it had to be a business that would benefit society. It also had to have considerable potential for growth and development. Finally, from his experience with debt collection came a determination that his business would be one which could operate on a payment-in-advance basis. He even rejected the practice of paying with post-dated cheques. His resolve to go against prevalent business custom resulted from the knowledge gained during and after the war when, as a young adolescent, he had come to realize that no established value should remain unquestioned.

It has been recognized that homogeneity in thought and in behaviour is a characteristic feature of the Japanese, encouraged from birth while 'abnormal' behaviour is actively discouraged. *Deru kugi wa utareu* or 'the nail that sticks out should be hammered down', is a well-known phrase in Japan. For success in business, however, conformity is not necessarily an asset. What is required is uniqueness and innovation. Iida and Toda were searching for uniqueness in their business venture. It was a search filled with uncertainty, the beginning of an expedition into uncharted waters.

They first contemplated the possibility of opening a bowling alley. Japan's first bowling alley had opened in Tokyo in 1952 and subsequently the sport had gained considerable popularity throughout the country, particularly among young urbanite workers and students. In addition to providing a site for physical recreation, most bowling alleys offered tearoom/restaurant facilities where young people could meet and socialize. While the bowling alley business had already been introduced to Japan, the tremendous popularity the game and its social function were to enjoy throughout the nation had not yet occurred. The sizeable outlay of capital required to establish a bowling alley made entry into this field difficult for Iida. At the same time, Iida felt that any business venture in which the primary entry requirement was money would not provide him with the challenge he was seeking.

In the ensuing decade entrepreneurs invested heavily in bowling centres. By the time the bowling craze reached its peak in 1972 there were nearly 4,000 centres nationwide. Over the next four years, however, competition among the centres and the decline in the popularity of the sport combined to bring about the closure of 70 per cent of these centres, many going bankrupt. In retrospect, Iida is thankful he did not enter this field.

Another venture briefly contemplated was the establishment of a consumers' cooperative. What he gave greatest thought to, however, was the establishment of a mail order business. For several years after the war there had been very few consumer goods available for purchase in Japan and certainly no luxury items. During this period Iida had from time to time encountered a Sears Roebuck catalogue: 'The catalogue held so many dreams for us during those years of deprivation. There was nothing similar in Japan.'

Had Iida followed through with this idea success would have been hard won. It was an idea whose time had not yet come. Mail order shopping never did catch on in Japan the way that it did in North America. It was not until the late eighties that catalogues began to carve out a niche for themselves. High shop rents, limited display space, and a complicated, multilayered distribution system have made the introduction of new products to the market via conventional channels difficult and expensive. Facilitated by the emergence of reliable, fast, economical parcel delivery services, catalogue marketing circumvents the distribution network and eliminates the need for expensive shop space. The consumer benefits through savings in money and in time spent shopping.

Many ideas had presented themselves to Iida as avenues for entrepreneurial venture. What was crucial in respect to 'opportunity' was his evaluation of these ideas.[4] In the autumn of 1961, while he was in the process of seriously investigating the establishment of a mail order business, a university friend of Iida's recently returned from a trip to Europe told him of the existence of foreign security services firms. As Iida listened he began to realize that this could be the type of business for which he had been searching.

It was an innovation. As Rogers points out, it is not the 'objective' newness of an idea – measurable in time elapsed since its initial use or discovery – that is the issue. What matters in respect to human behaviour is the perception of newness by the individual. It is this perception of newness that determines his reaction to it.[5] Security service as a business opportunity was a concept new to Iida. The innovation here was not the designing of the 'system' but the realization that there was a need for a system, that is, security in all its forms.

The entrepreneur, like the artist, is an originator capable of seeing the world differently from most of its inhabitants. The surroundings perceived by the original mind are ripe with suggestion. All perception entails interpretation, the injection of meaning into sense-impressions, their comparison with remembered experiences, the recognition of the possibilities they hold. The creative individual is characterized by these same but enormously enriched and intensified thought processes.[6] Iida needed neither elaboration nor a detailed description of security activities in foreign countries. He could visualize the type of business he wished to establish simply on the basis that such a business existed elsewhere.

Japan's public police system, patterned on the German model, had been established in 1874, soon after the Meiji Restoration. By the time World War II broke out, Japan had become a police state. The absolute control exercised by the national government over the strongly centralized police bureaucracy made it possible for the military 'to spread a network of political espionage, suppress freedom of speech, of assembly, and even of thought, and by the means of tyrannical oppression, to degrade the dignity of the individual.'[7]

After the war, on the orders of GHQ (General Headquarters), fundamental changes were introduced. The police system was reorganized into a National Rural Police Force and approximately 2,010 independent municipal police forces. A National Public Safety Commission was established to administer the National Rural Police Force and the Fire Defence Board. Each municipality was responsible for peace-keeping activities within its boundaries through its own police force, independent of the central government. Each municipal police force was headed by a chief appointed (and removed) by a three-member civilian commission appointed for a fixed term by the mayor with the consent of the local assembly. There was no channel of command between the National Rural Police Force and the municipal police forces although in the interest of mutual assistance, liaison, and coordination, technical channels of communication were effectively maintained.

Police responsibilities were limited to 1) prevention of crime, 2) suppression of criminal activities, 3) detection and apprehension of criminals or violators of the criminal laws, 4) protection of life and freedom, 5) preservation of law and order, 6) control of traffic and traffic safety, and 7) serving warrants, subpoenas, and other court instruments. Extraneous activities, performed by the police prior to the reorganization, were transferred to other government agencies with the removal of the Sanitation Section, Fire Department, Census Bureau, and Economic Controls from the police department. Special guard duties such as railway guard, bank guard, occupational guard, and so on were excluded from permissible police activities.[8]

When required, such guard duties were filled at the corporate level with individual organizations providing in-house personnel. Historically, feudal lords and, later, rich merchants had hired body guards to protect both their persons and wealth, but organized private security services did not exist in modern Japan. The absence of the necessary know-how within the country provided an added dimension of excitement and challenge for Iida.

Iida was aware of the need for on-site security at special events, the emphasis being placed on crowd control. Over the centuries festivals and special religious celebrations at temples and shrines drew large crowds. Promoters of these events focused on attracting large numbers of participants while holding little concern for their safety. On New Year's Day, 1956, 15,000 people congregated at Yahiko Shrine in Niigata prefecture. Despite the fact that thirty-six city police officers were present along with ten internal security men, 124 people died and ninety-four were injured as the result of the extreme congestion. Festivals occur regularly throughout Japan attracting huge crowds. Every year there are many injuries and fatalities.[9]

Another activity that drew large crowds particularly in the early sixties was musical performances. Here, too, one accident could cause a chain reaction that would lead to many injuries and deaths. On 2 March 1960, 6,000 people gathered at Yokohama city sports arena for a concert. Despite the fact that nineteen policemen were in attendance twelve people died and many more were injured. These and other similar events led to a public outcry and a demand for better crowd control.[10]

Municipal police forces throughout Japan were facing a problem shared by their counterparts elsewhere – excessive fragmentation of slender resources in the attempt to satisfy a variety of constituent demands. The police argued that it was up to the events' organizers to provide the primary security service. Although he had not yet crystallized the objectives for his proposed company, Iida felt that this was not an area of particular interest to him. What he envisaged was much broader in scope.

In exploring the concept of security services it was important to Iida that his security business should satisfy all three of his aforementioned conditions. Certainly, it would be socially beneficial to participate in the provision of safety measures which would help satisfy the basic human need for a safe environment. For a healthy society to develop, security is essential. As there was no precedent for the security business in Japan, its success would be that of the entrepreneur's conceptualization, with its effective and economic actualization. It seemed to Iida that the business opportunities it would provide would be unlimited. For a security business to meet his third criteria, payment-in-advance, was more difficult; nevertheless, he felt he would find a way to make it work.

Iida had worked for his father for five years. Having settled on the type of business venture he wanted to get into, he felt the time had come to leave the family firm. When he spoke to his father of his resolve the older man was dumbfounded by his son's choice of a business, as yet unheard of in Japan. Opposed to the idea of his youngest son involving himself in a line of work that did not yet exist in the country, he ordered Iida to remain where he was. Iida complied. Possibly thinking that Iida's desire to strike out on his own might have grown out of the unpleasant conditions under which he was forced to work and his lack of status within the company, Iida's father made him a company executive and member of the Board of Directors. More likely the senior Iida was motivated by the knowledge that sooner or later his son would leave the family business and his desire to complete the education process he had begun when he insisted that Iida work for Okanaga at the outset of his career.

While Iida was not yet ready to challenge openly his father's wishes, his improved position in the company did not lessen his determination to found his own security firm. With Toda's help, he continued to examine the feasibility of its establishment. Knowing of the existence of businesses which sold security services was sufficient to kindle Iida's imagination. He decided it was unnecessary to go to Europe in order to see what form of security services were in operation; becoming too familiar with the operational details of existing security services was to be avoided, at least initially. Iida feared he might copy their practices rather than initiate a service appropriate to the Japanese market.

Iida is an independent thinker. His propensity to act on his own has been manifested over the years by his refusal to bring management consultants into the decision making process. He prefers to reflect on the essence of a problem and reach his decision independent of external influence, which would, he feels, disrupt his concentration. He likens the development of his business to painting on a blank canvas. In the late sixties, writing in a corporate newsletter, he stated:

As an artist thinks about the essence of the picture he will paint as he faces a blank canvas, so I think about the essence – the ultimate mission and future direction – of my corporation. When I concentrate on this point the issue is not how much money I can make but rather what my business should be in order to provide the best service possible for the betterment of society. My management endeavour has been the continuous search for the resolution of this issue.

This statement reflects Iida's entrepreneurial mindset. For Iida, 'money is not the incentive to effort but rather the measure of its success.'[11]

Several months after the initial confrontation Iida approached his father again. The older man remained unyielding. If he left the family business his father would disown him, not in the prewar sense of erasing his name from the family register but in the postwar tradition of refusing his son help of any kind. Iida's brothers opposed his intentions as well, feeling that his business plan held little potential. But Iida was not deterred. This time he was ready to take the consequences of his decision.

He had never intended to rely financially on his father. As well, he felt that no matter how thoroughly he explained his business intentions, his father would never really understand his objectives. Nor would his father's years of experience in the sake wholesale business provide the expertise he would require in his own endeavour. He resigned from Okanaga. Subsequently he and his wife and son moved from their quarters on the family property to a rented house in Fuchu city on the western outskirts of Tokyo, about 25 kilometres from Nihonbashi.

Reflecting on this time, Iida feels that his father's decision to disown him was made not out of stubbornness or meanness but rather in an effort to strengthen Iida's independence. The severing of the relationship forced Iida to stand on his own feet. There was nothing to fall back on, no line of retreat. This was his Rubicon.

Iida's brother, Tamotsu, recalls his feelings at the time. He wondered if Iida really could make a living in his new venture. He knew that their father appeared to be against Iida's decision and yet he was uncertain of the strength of his opposition. He had always encouraged his sons to establish their own businesses. If he helped Iida financially perhaps his youngest son would develop a false sense of well-being. Whatever the father's motives, Tamotsu was encouraged by his brother's decision to pursue his own independence.

However, Tamotsu's separation from the family business did not come until 1969, a year after his father's death from a cerebral thrombosis at the age of 77. This was seven years after Iida's independence and twelve years after that of the third son, Susumu. Before Tamotsu opened a pub in the Ikebukuro district of Tokyo, the first of over sixty in what is now the Tengu chain, he spent a year travelling throughout Japan and Europe gathering ideas. Tamotsu had learned from his father that one key to success lay in providing quality items at the lowest possible cost to the customer. Success in this regard hinged on the effective acquisition of supplies. Tamotsu's independence was so long delayed because his father, recognizing his impulsive nature, restrained him, waiting for him to mature.

Eventually three of the four brothers who worked for Okanaga struck out on their own; Susumu in 1954 at the age of twenty-six, Iida in 1962 at the age of twenty-nine, and Tamotsu in 1969 at the age of forty. It has been pointed out that 'the decision to become an entrepreneur is most likely to be made between the ages of twenty-five and forty. . . It is a period in which the entrepreneur is most likely to be able to make a decision.' [12] While the ages of Iida and his brothers at the time they became entrepreneurs fit within this broad framework, their father was a significant influence in the timing of their individual decisions.

Following the age old Japanese tradition and his father's wishes, Hiroshi, the eldest son, inherited the family business. As a graduate of Tokyo University, Japan's most prestigious institution of higher learning, with a degree in economics, job opportunities in both the public and private sector would have been readily available to him. But familial duty came first. Within the confines of the inherited environment, Hiroshi took innovative steps, as had been advocated by the senior Iida throughout his life.

He was dissatisfied with the availability of only a handful of brands of sake in major retail outlets, feeling that quality was being sacrificed for the lower prices under the system of bulk purchasing. This practice was lowering the quality of sake throughout Japan, he felt, eliminating the small producers of quality products. Realizing that several brands of regional sakes were available but relatively unexposed to consumers due to the absence of the necessary distribution channels, he established a new distribution network and promotional system to bring these sakes into the limelight.

He selected twelve sake producers located in various parts of Japan and began distributing their products to interested retailers in 1975. Among the some 5,000 sake brands in Japan he chose two dozen which were not linked to major distribution channels. He met with the chief brewer and the president of each of these, studied the brewing operation, and personally sampled each of the products before deciding on the twelve brands Okanaga would distribute. The final step in the selection process could only be carried out after Hiroshi had overcome an allergy to sake that resulted in an outbreak of hives and a bad cold each time he consumed the liquor.

Atsushi, the brother two years Iida's senior, suffered from ill health most of his childhood and young adult years. Despite this setback, he now owns several establishments, including a tangerine grove in Shizuoka prefecture where he lives and a number of rental properties. He is the only brother to have moved away from Tokyo. All five men have done very well in their endeavours.

Having successfully introduced a new type of business to Japan, Iida is now frequently asked to speak on starting a new business. He says:

> Addressing this topic is difficult as there are many variables, for example, economic conditions; business environment; family situation. The opportunities to start a new business are always numerous. Whether or not one can find the 'right' business depends on individual drive and determination. If one simply waits for the right opportunity to come along, it never will. To succeed, one needs a burning desire to be independent. Fired by this intense drive one must examine the world in which one lives. It took me several years to find the business that was best for me but it was worth the effort I put into the search.
>
> I had no specialized knowledge and lacked a deep understanding of the complex nature of the business world. But in retrospect I think this was to my advantage. Had I been a specialist, I would likely have confined my search for business opportunities to my specialized field and ignored the possibilities in other areas.
>
> I convinced myself that there was no doubt I would succeed in the challenge that confronted me. The security business was a virgin field in Japan with unlimited potential for growth. If one succeeds in a business venture that no one else has challenged the new company will immediately become the top corporation in the field. Belief in oneself is essential for success. I am often asked how I have been able to sell security in a country that is among the safest in the world. In retrospect, it is possible to say you can sell almost anything in Japan.
>
> Recently in Japan there has been an upsurge in the emergence

of new businesses. Small, new enterprises are entering into ventures that the larger, established corporations never dreamed of. I welcome this. It will revitalize society in many ways. Today's young people excel in sensitivity and intelligence. It is heartening to see the enthusiasm and the creativity with which they challenge new business ventures. In choosing these ventures it is important that these young entrepreneurs not only capitalize on their intelligence and sensitivity but also leave aside their preconceived ideas and look with fresh, clear eyes at the direction in which society is moving.

To actualize a new venture a certain lightheartedness or cheerful liveliness is essential on the part of the entrepreneur. With this kind of attitude it is possible to overcome apparent limitations and take great strides forward. This in turn fosters lightheartedness and the momentum builds propelling the entrepreneurial mission.

In answering the questions 'What should my business be?' and 'How should I proceed in developing the business?', Iida feels it has been to his advantage that he is not a technical specialist. His greatest strength lies in systems design. Without detailed knowledge of the available hardware he is not restricted in his vision by preconceived ideas of technical limitations. The foundation for excellent systems design is logical thinking, not mechanical or engineering ability, he argues. For Iida, constructing a system is not an arduous task. The training he received in systematic thinking through his studies and his involvement in sports at university have served him well.

However, logical thinking, in itself, is insufficient. It must be accompanied by the ability to predict the future. Says Iida:

Corporate leaders are ordinary human beings. Predicting their own future is a nearly impossible task, never mind the future of the corporation. Nevertheless corporate management is not like the Olympic Games where participation itself has some meaning. If one simply participates the corporation will soon be overtaken by its rivals. The company must move faster, jump higher, and grow stronger than its competitors. Within this context reading the future becomes top management's most important function. If this is done even a little, then the corporation can prepare for future growth and can strengthen its weak aspects.

4 Originating an industry

PREPARATION

In the spring of 1962, in order to carry out the preparatory work necessary for the establishment of Japan's first security company, Iida rented a small one-room office in the Kudan district of Tokyo. He and Toda were its only occupants.

The scenery from the office window was beautiful. It overlooked the garden of the Indian Embassy and Chidorigafuchi, a moat at the Imperial Palace. As the two men watched the flurry of cherry blossom petals fall from the trees in the garden their dreams of success expanded. The apparently infinite number of petals inspired thoughts of countless numbers of possibilities for future expansion. Toda recalls this period:

> At this time we collected reference materials, planned for the future, and dreamed about what we would become. The struggle to make the company a reality was long and hard but it was also thrilling. When I close my eyes I can still see the cherry blossoms in bloom outside the window of our first office.

The preparatory period was, for Iida and Toda, a time to study and to make concrete plans for the future direction of their company. It was a time when Iida did a lot of reading. One area of interest at this time was systems design as he wanted to understand the principles involved:

> Understanding principles and theory is important. I read eclectically always selecting material that is interesting to me. Nowadays I do not read business books, especially those of the how-to variety. One must design one's own business oneself; there is no room for other people's designs thus reading about them is unnecessary.

On the practical side, Iida and Toda collected and analysed statistical data. Although Japan was, and still is considered to be a 'safe' country regarding crime, statistics supported their belief that there was a definite need for security services.[1] In the five years from 1956 to 1960, according to reports of the Police Agency, the country experienced 13,152 murders, 26,083 robberies involving violence to the victims, 26,340 incidents of rape, and 8,252 cases of arson.[2]

Beginning in 1950 an important task of policemen on the beat – as they carried out periodic checks of the premises in their area – was to keep in close contact with the families and commercial establishments within their patrol territories. The purpose was to ensure neighbourhood safety and to exchange information regarding the prevention of crimes and accidents and preparation for disasters. However, societal conditions such as overcrowding, lack of zoning resulting in extreme diversity of activity in a limited space, and the desire for anonymity made the conduct of police periodic checks difficult and a financial burden to the taxpayer. This resulted in a decline in quantity and quality of service to individual societal members.[3] In 1962 the Tokyo metropolitan police force numbered 27,356 officers with 1,273 neighbourhood police boxes, and 250 patrol cars.[4] A niche emerged for the services of a private security firm.

As well, public awareness of the need for crime prevention measures was increasing. In the postwar turmoil loosely structured regional crime prevention associations, *Bohan Kyokai*, emerged throughout the country in response to local crime. In the spring of 1962 the National Federation of Crime Prevention Associations (*Zenkoku Bohan Kyokai Rengo-kai*) was formed. Its branches are closely linked to local police stations across Japan. Its 'grassroots' activities include local patrols, mutual protection information exchange among individuals and between individuals and police, security public relations, organizing crime prevention seminars, and making recommendations for improvement to the police. Households nationwide have volunteered and been selected for designation as communication points for the Federation.[5]

The more Iida and Toda contemplated the existing situation and the inherent societal need for the professional service they were proposing, the more certain they became of their success. Excited at the prospect of his new venture, Iida was eager to discuss it with his friends. Shinichiro Osada was one with whom he shared his excitement, asking Osada to guess what kind of business he had decided on. 'The extermination of white ants', was Osada's reply, not an unlikely choice as ant eradication was a burgeoning industry in Japan in the early sixties. When Iida enlightened him with details of his plans to establish

Japan's first security firm Osada felt that the concept had militaristic overtones. Certainly he never imagined that one day he himself would be engaged in this business with Iida. Osada, now a senior executive director of SECOM comments, 'I did not think that such a militaristic endeavour was well suited to Iida. In retrospect, though, I realize that Iida's image of a security corporation was not dependent solely upon regimented manpower.'

Despite the fact that the Japanese economy had strengthened remarkably since the end of the war and was about to soar to unheard of heights, the Japanese recognition of the necessity of security was low. With regard to business offices and factories, what security existed was provided either by an employee who would sleep at the site or, more commonly, by a retiree hired to check periodically on the premises during the night. On the latter's elderly shoulders rested an enormous responsibility. That break-ins and theft were infrequent was due more to coincidence than to tight security. The relatively low crime rate in Japan caused most industrialists to give a low priority to security concerns.

Historically, internal wars, conspiracies, thievery, arson, and natural calamities such as earthquakes, tidal waves, volcanic eruptions, and typhoons abounded in Japan making the country unsafe for many, regardless of status and occupation. To cope with the uncertainties of existence on earth, the Buddhist philosophy of predetermined destiny became integrated into the Japanese thought process. One's fate was perceived to be sealed. Nothing could change the inevitable so precautions were deemed to be of little importance. Viewed in this context it is perhaps not surprising that even in the early sixties few corporations were willing to pay for a high level of security.

Iida, however, felt that even though societal needs for security had not been given prominent historical/cultural expression, if he could provide a high quality, low cost, professional service he could capitalize on the gradually emerging societal changes and generate a demand for his security service. Rising property values, life in general involving more risks, the growing need for individuals to take responsibility for their own self-protection, and ensuring institutional security through the protection of raw materials, manufacturing parts, corporate documents, and information provided a raison d'être for the security industry.

Now that Iida had office space, he considered what other basic equipment he would need to launch his business. Guns were not required; in fact, they were prohibited by law to all but the police and licensed hunters. It was only after the end of World War II that all police officers were issued firearms. Until then their distribution had been selective. In 1949, under the direction of GHQ, guns were borrowed from the American military forces for a period of five years and distributed to all police. They were formally given to Japan in 1955. At this time 90 per cent of the guns carried by Japanese police were American made, left over from the war. Their size and weight varied but generally they were too large for the Japanese and inconvenient to use because of differences in bullet size and operational methods. It was not until 1960, as a result of postwar research, that Japanese-made guns were introduced for police use.[6]

Iida would need to supply his guards with uniforms, nightsticks, and an automobile. In addition he would need sufficient funds to cover salaries and rent. To meet these expenses Iida and Toda calculated that they would need four million yen at their disposal when they established the company. The financial resources available were far short of this amount. Iida had 500,000 yen, saved from his previous employment, to put towards the establishment of his business.

When Iida's two brothers struck out on their own in the business world their father, approving of their actions, backed them financially. Iida's independence, however, was against his father's wishes; financial assistance was out of the question. The only avenue open to Iida was to seek bank loans. Iida's father was a well-established and highly respected large-scale sake wholesaler. Had he been the guarantor, securing a loan would have been a simple matter but Iida was determined to find his own way.

He approached a number of financial institutions but in every instance was rejected before he had a chance fully to present his case on the grounds that the business he was proposing was a service industry, that the concept of a security business was not well understood in Japan, and that Iida was without collateral of any kind.

Iida was proposing an innovative business venture. Yet innovation, by definition, is full of uncertainty and is often beyond the range of comprehension of those not involved. Its features and, more significantly, its economic potential, are seldom as obvious to a prospective source of financing as they are to the innovator. Viewed within this context the banks' response was not surprising. No other alternatives presented themselves so Iida prepared more detailed future business plans and financial projections and visited more loan officers, explaining his proposal enthusiastically.

Promoting his intentions in this way was a valuable learning experience, forcing Iida to be 'realistic' and precise in his objectives and providing him with knowledge of the functioning of financial institutions and the nature of their decision-making processes. Financial institutions function according to a myriad of rules and regulations but, none the less, they are run by people. Even within their rigid environment, Iida came to understand, there is the opportunity for human interaction which will sometimes produce unexpected results.

While collateral was advantageous to loan procurement, Iida learned that its presence was not always necessary. Entrepreneurial vigour coupled with commitment to one's prospective business and the presence of well thought-out business plans figured in a bank's decision-making process as well. As Iida became more diligent in his approach to financial institutions, they became more responsive and eventually a credit union agreed to provide Iida with a loan of 1.5 million yen. Banks are often perceived as a source of financial assistance only for those who already have 'wealth'. Through his experience Iida learned that they are also capable of lending to those without assets. Even with the loan Iida still needed an additional two million yen. The remaining start-up money came from an unexpected source.

Through his research he learned of the existence of Ligue Internationale des Sociétés de Surveillance, an international security federation to which a representative company from each of a dozen countries belonged. Established in 1934 the Ligue had four charter members – Switzerland, Austria, Denmark, and Sweden. Its mandate is to propagate the mission and purpose of the private security firm across national boundaries. Through discussion of possible solutions to problems confronting the security industry in member countries and the exchange of information regarding security technology and operative methods, the Ligue promotes the mutual assistance of its members while not inhibiting free and fair competition among them, and contributes to society at large through the prosperity of the security industry.

In order to represent one's country in the Ligue, a company has to have been in existence for at least five years, has to meet a total sales minimum, and has to be free of political connections. It is usually a country's largest security firm that is its representative. By the time of its fiftieth anniversary in 1984 the Ligue had twenty-four member companies representing twenty-two countries. The Ligue is headquartered in Bern, Switzerland and meets biannually to elect its officers.

Iida solicited the support of the Ligue's chairman, Erik Philip Sorensen, managing director of DeForenede Vagtselskaber,

Copenhagen. Philip Sorensen was one of four founding members of the Ligue, the others being Moritz Steiner, director of Bewachungsgesellshaft der Industrie, Vienna; Jacob Spreng, general manager of Securitas, Switzerland; and M.C.L. David, lawyer at the Supreme Court, Copenhagen.

Iida wrote to Philip Sorensen of his intention to found Japan's first security corporation. In his reply, Philip Sorensen expressed interest in establishing a security firm in Southeast Asia. He pointed out to Iida that, in any business, the founder's morale and attitude determine the direction the business will take. A security business, in particular, strongly influences the society in which it exists. He urged Iida to postpone the establishment of his company until he could visit Japan and discuss the situation more fully.

Two months later Philip Sorensen made the trip to Japan. Confident that Iida's plan would succeed, he offered to invest 2,010,000 yen in the company and so become the majority shareholder. Iida was impatient to launch his business and somewhat naive regarding financial matters. He accepted Philip Sorensen's offer. Iida and the credit union together put up the remaining 1,990,000 yen and the company was born. It would be eight years before Philip Sorensen would agree to Iida's desire to buy him out. On Saturday 7 July 1962 Iida registered the company under the name Nihon Keibi Hosho Kabushiki Kaisha or Japan Security Guard Corporation.

The term *keibi hosho*, now widely used in the security service industry in Japan, was coined by Iida. *Keibi* means 'defence' or 'guard' and *hosho* is translated by 'guarantee' or 'security'. From the modern Japanese point of view, the name is unpolished. Iida chose it thirty years ago to reflect accurately the nature of the specific security service he was providing. He also gave the company the English name Security Patrol Company. Twenty-one years later the name Nihon Keibi Hosho would be changed to SECOM (Security Communication).

Iida chose 7 July because it is a significant day in Japan marking the annual *Tanabata* festival. According to legend the Star Weaver (Vega) and her lover the Cowherd (Altair), separated the rest of the year by *Ama-no-gawa* (Heavenly River) or the Milky Way, meet once a year on the evening of 7 July provided that it does not rain. Rain, it is believed, will flood the Milky Way and prevent the lovers from being reunited. But it did rain on *Tanabata* day in 1962, not surprisingly perhaps as 7 July falls in Japan's rainy season. In fact, on *Tanabata* night and throughout the following day torrential rains pelted southwestern Japan causing floods and close to 300 reported landslides. Transportation was paralysed and the number of dead or missing reached fifty-six. The weather seemed indicative of the difficulties that lay ahead for Iida.

The economic climate, however, was in his favour. By 1962 the primary, secondary, and tertiary sectors were nearly equal in terms of the percentage of the labour force they employed, however, the tertiary sector was beginning to pull ahead of the other two. The economy was rapidly gaining strength and many new service industries were emerging. Iida's business was one of many established in the early sixties. This in itself, however, did not guarantee success. Iida's eventual hard won success resulted from his ability to plot his course in advance of societal trends.

After the company was officially registered, one of the first tasks undertaken was to design a corporate logo. Iida selected a stylized owl, popular with European security firms, to be the central feature of the logo. As the bird of wisdom and of darkness and death, the owl is an ambivalent symbol. Different symbolic meanings have been given to the bird by different cultural groups. In Japan it is esteemed for its faithfulness, always present in the forest. In his selection of a corporate logo, Iida associated the bird with constant vigilance especially throughout the night and chose to focus on the Graeco-Roman association of the owl with wisdom. In this tradition, the bird was sacred to the goddess Athene/Minerva.

Above the owl was the Latin phrase *vigilamus dum dormitis* meaning 'vigilance while you sleep'. Two crossed keys symbolizing trust were below the owl and beneath them were the letters SP standing for Security Patrol. This logo remained in use for eleven years, appearing in a variety of places including the corporate flag, patrol cars, letterhead, guards' uniforms, and business cards. While Iida deliberately avoided modelling the content of his business activities on European security firms, he felt there was no harm in adopting and adapting their prevalent symbols in order to give his company a 'modern' image.

By the time the company became a reality Iida had moved to a new one-room office. Located in the Shiba Park district of Tokyo, it was on the second floor of a building owned by SKF, a Swedish ball-bearing manufacturer. The 17 square metre space was shared by Iida, Toda, a secretary, and the two men Iida had hired as guards. There was much to be done and to be learned through trial and error. Iida and Toda devoted time to developing the terminology essential for operating a security firm, drawing up fee schedules and contract forms, and preparing for training guards. Their youth was an asset in

2 Toda and Iida (back row left) with a Swedish security representative and Nihon Keibi Hosho's staff in front of a patrol car, 1962.

work that was both mentally and physically taxing.

A number of men had responded to Iida's newspaper advertisements for personnel to work as security guards but Iida felt the quality of applicants he was seeking was lacking. Whomever Iida and Toda selected to be their first employees would eventually supervise and train future guards. Seeking men who were virile, honest, and yet humble, and wanting to make the best possible objective decision, Iida enlisted the help of the late Hitoshi Aiba, a psychology professor at Waseda University in Tokyo. Aiba introduced Iida to the Guilford–Martin Personnel Inventory, a screening test to identify the applicant's objectivity, agreeableness, and cooperativeness developed between 1943 and 1946 by J. P. Guilford, a psychology professor at Stanford University in conjunction with the US Air Force.

The first employees hired were Japanese who had been employed after the war as security guards at American military bases near Tokyo. They had been trained by the Americans in regimented security drill activities. Employee selection was one problem. Orienting them to corporate philosophy was another. Iida and Toda discussed professional ethics and social responsibility, instructing the men in 'sincerity', 'responsibility', 'alertness', and 'service'. These were the cornerstones of the corporate operation and the words appeared on the office wall and, later, on company advertising. These first employees would later be used to train additional guards. With these employees in place, the next task was to secure contracts.

INITIAL STRUGGLE

Before a business becomes socially recognized it takes courage and determination to hold fast to rigid conditions of business operation with respect to customers. On the other hand, to be able to set new business practices is the privilege of the first company to enter a new field.

Going against prevalent business practice and unswayed by his familial responsibilities, Iida stood firm in his resolve to operate his company on a payment-in-advance basis. Contracts, he decided, would be for a period of two years with a penalty for early termination. Payment would be due every three months, in advance of the provision of service. In addition to the avoidance of payment collection problems, payment-in-advance, Iida felt, would provide his customers with a guarantee that the contracted service would indeed be delivered. This view, however, was not widespread. More commonly, customers wished to delay payment in order to maintain a position of strength and some power of negotiation should the promised service prove unsatisfactory. Friends and business acquaintances opposed the severity of his approach and advised him repeatedly that he would not be able to attract customers.

When he left Okanaga, Iida's father had forbidden him to use the company name to promote his own business. Now that the time to market his security service had arrived, Iida made a conscious decision not to use connections of any kind. It is customary in Japan to conduct almost any type of business with the help of 'go-betweens' – mutual acquaintances who facilitate business dealings through the introduction of one party to the other. Iida wanted his customers to understand fully the merits of his service and base their decision to purchase it on their own evaluation, not on the recommendation of a third party. As well, he feared that business contracts secured through connections would, under the Japanese sense of *giri*, often expressed as 'duty' or 'honour', unnecessarily obligate him.

The first contract was a long time coming. Iida's innovation – security service provided by an independent firm – was not compatible with the existing values and cultural norms of the potential adaptors. In the early 1960s the Japanese had no past experience with such an externally provided service. Of course, had the concept fitted exactly with current practice, innovation would have been absent.[7] The service

Iida was proposing was 'revolutionary'. There was no ready-made market; it had to be cultivated. The service offered was 'persuasive merchandise'. Iida recalls:

> We didn't know precisely who our customers would be. Toda and I simply went out and knocked on doors. We were very enthusiastic and there was considerable interest on the part of our prospective customers in what the company was all about. Sometimes the person with whom I was speaking would listen attentively but when I finished the response was similar time after time: 'How old are you?'[8] 'Oh, you have thought of a very interesting type of business. How did you come up with such an idea?' No one signed a contract. This initial market entry effort was extremely time consuming and its sales efficiency was zero. We were not impatient, though, as our belief in our eventual success was strong.

Remembering their initial struggle Toda comments:

> By the early sixties television was well diffused throughout Japan. While we hadn't yet reached the point of 'one in every home' most people had a fair amount of exposure to it. When I went knocking on doors trying to sell Nihon Keibi Hosho's service many were closed in my face. From the people who did listen to what I had to say the response was often, 'You have been watching too many American television programmes. We have no need of a security service here.'

The widely accepted view was that protection of corporate property was the responsibility of the corporation. The idea of entrusting a 'stranger' with a key and employing him to guard the premises was unheard of. Even without contracts, Iida calculated that the four million yen the partners had invested would keep the company afloat about ten months. Three months passed and still the first contract had not been signed. There were frequent questions about the contract and payment policies. Potential customers were puzzled about why they should hire 'outsiders' to protect their property and pay in advance for a service the quality of which they knew nothing about. Some were flabbergasted; others were angry. Some said they would sign a contract if they could pay for the service after they had received it. Others offered to pay in advance if they could first try the service for a month free of charge. Still others wanted to pay with post-dated cheques. Loans, instalment payments, and credit were well understood but Japanese consumers had little knowledge of prepayment systems.

Iida would not alter his stance. Standing firm was, he felt, a test. Clients agreeing to pay in advance for his security service would testify to the need for a security industry to exist in Japan. Nor could Iida afford to provide security in advance of payment. True, if he had agreed to do this there would likely have been no shortage of contracts. Serious problems with cash flow might have developed, however, to the point where expansion (additional contracts) might have had to be curtailed due to insufficient funds to meet operational expenses and cover capital investment. In any event, to abandon his prepayment scheme would have meant that Iida would have had insufficient control over finances to visualize and plot the company's systematic progress. It was a critical moment in the life of Nihon Keibi Hosho. Then on 24 October 1962, three and a half months after establishment, the first contract was signed with a travel agency specializing in international travel. The agreement was that a security guard would check the premises at regular intervals during the night. Iida and Toda were jubilant.

During the next five months about ten contracts were signed. Iida's initial success was due, in part, to the booming economy. Prime Minister Hayato Ikeda, in office from 1960 through 1964, had embarked on a bold programme of economic expansion, envisioning a doubling of national income by the end of the decade. To achieve this goal would require a minimum annual growth rate of 7.2 per cent. The target average growth rate was set at 7.8 per cent. One focus of the national economic policy was the stimulation of government spending to improve roads, harbours, water supply, and sewage disposal, areas heretofore seen as hindrances to effective economic growth. Another focus was human resource development through the expansion of educational and training opportunities in order to promote science and technology, to overcome shortages of skilled labour, and to prepare Japan for further trade liberalization. The third focus was the elimination of regional discrepancies and discrepancies between large- and small-sized industries, among others.

Ikeda's plan brought striking results. The annual national growth rate for the decade averaged 11.6 per cent, well in excess of the targeted 7.8 per cent. The industrial infrastructure was well advanced and the enrollment rate at institutions of higher education had increased dramatically. As well, there was some reduction in regional and industrial hierarchical discrepancies. However, there were some negative results, too. Rising inflation and widespread environmental pollution were two major costs of economic success.[9]

With factories and businesses growing rapidly in number, construction was in progress everywhere, and there was a corresponding increase in the potential need for security. In accordance with the rising demand for the service Nihon Keibi Hosho offered, Iida hired and trained additional guards. Selection continued to be based on the Guilford–Martin Personnel Inventory.

In retrospect, Iida feels the test did not lead to the selection of the type of employee he wanted for his company. The first fifty or sixty men hired were honest, sincere, and cooperative but there was too much similarity among them. Generally, they lacked initiative and inquiring minds. This was not surprising given that the test was designed to seek out agreeableness and cooperativeness, character-istics commonly shared by the majority of Japanese for whom conformity is a societal virtue. Realizing that the test was not producing what he considered to be ideal employees, Iida began to select those men who, according to the test, varied slightly from the 'standard personality'. A certain amount of personality variation among employees, he felt, was essential for organizational vitality.

> At the outset I was nervous and apprehensive. I was striving to build an ideal corporation. Because it was other people's property my guards were responsible for, I tried to select employees who would not make mistakes. What I ended up with were men who were obedient and sincere but who lacked vitality. There was no competitive spirit and they lacked an eagerness to strive to improve themselves.

> When one is dealing with people as homogeneous as the Japa-nese, even when one seeks out their differences the results may tend more to the similar than to the diverse. Nevertheless, I feel it is necessary to select as employees people who are different from each other but complementary to the group as a whole. The result should be akin to an orchestra where each instrument has a unique sound.

As Nihon Keibi Hosho's reputation spread the number of employment applicants increased. In order to select the best from among them Iida actively participated in the hiring process. He placed more faith in the interviews than in the written tests. While he was interviewing a candidate he would observe the person's eyes looking for a sparkle indicative of enthusiasm and a zest for living. Iida continued to be directly involved in hiring decisions for the first fifteen years of the company's existence casting the deciding vote in the decision on some 12,000 employees. A day of interviews would leave him exhausted, but he was firm in his belief that hiring the right people was the key to the development of a dynamic corporate culture that would trigger innovative developments.

Guards were hired and uniforms and equipment purchased only as they were needed. Buying-on-demand is still the usual company practice today. SECOM does not operate on a budget system in the strict sense of the word, thereby making it easier to be innovative, to abandon the old and the obsolete. At the annual meeting of the Board of Directors approval is given to a budget for the coming year but the various sections of the company are not made aware of their budget allocations. At SECOM the budgets for ongoing business and for innovative endeavours are conceptualized differently and handled separately. From the outset Iida has acted in the knowledge that a different application of budgets and budgetary control from that required for the maintenance of a business is demanded by corporate innovative strategy.

For ongoing business the issue, acknowledges Iida, 'is the accomplishment of whatever is required with the minimum expense'. Of innovative efforts, Iida says, 'If not making an innovative change would cost the company more in the long run through lost opportunity than making it, then the issue becomes: what are the maximum human and financial resources that can be devoted to the actualization of the innovative endeavour?' Iida encourages all his employees to participate to varying degrees in expense minimization and innovative endeavours. Each request for funds must be accompanied by justification for the expenditure and authorization is dependent upon Iida's perception of it as legitimate and necessary, regardless of size. In short 'buying-on-demand' has two dimensions: minimum investment for ongoing business and maximum investment for innovative projects.

In the spring of 1963 the company had to vacate the Shiba Park office. Iida settled on the most economical premises he could find – roof-top space slightly larger than what he was leaving in Ogawacho in the Kanda district of Tokyo. To reach it, it was necessary to take the elevator to the top floor of the building, climb the stairs to the roof, cross the roof and ascend the fire escape to the door of the 'office'. Adjacent to the room that housed the elevator machinery, the room was an addition to the building, intended originally as a storage area.

The setting seemed an unlikely one for a company with a 'bright' future. A prospective employee arriving for an interview would often change his mind before entering the premises. Aware of this Iida and Toda posted a notice which read, 'With a company that sells necessities there is no waste.' This statement did little to alter the situation but it did attract media attention.

Iida and Toda soon recognized the necessity of holding interviews in a different location – one large enough to allow a private conversation. One of the places they chose for this was a conference room in the Daiwa Bank in the Ueno district of Tokyo. Reminiscing about that time, Yoshinao Asado, one of Nihon Keibi Hosho's first guards, and now operation manager of Jastic Kansai[10] said,

> I had been a member of the Self-Defence Force and my duties included instructing in military facilities technology. After I left the Self-Defence Force I joined a private company. I found it very loosely run and I missed the regimentation of the Self-Defence Force. When I saw a small newspaper ad. for a position at Nihon Keibi Hosho I was attracted by the new term *keimushi*, or 'guard engineer' and the relatively high salary compared to other work available to someone with my qualifications. In response to my application the company sent me notification of the examination.

> I took a day off from my job and went to the Daiwa Bank conference centre for a written examination and interview. I was told I was one of seventy taking the exam out of a total of 200 applicants. In the morning, before the test, a uniformed examiner talked to us about the company and explained its future plan. My desire to be employed by the corporation intensified. The exam was followed in the afternoon by an interview with Mr Iida, Mr Toda, and Mr Matsumoto, the general manager. The decision was made that I would be hired provided my medical examination proved satisfactory. I was one of eleven successful candidates.

> A month and a half later, in September 1963, after tying up the loose ends at my other job, I joined the company. The first day I reported to the 'head office'. There I saw the examiner who had given the inspiring talk about the company and a female clerk. I was shocked by the size and state of the office and surprised that there weren't more people in it. In talking about the company the examiner had said that the employees would someday number 30,000. I assumed that it was already a reasonably large operation.

The office interior was no better than its exterior. One had to bend to get through the doorway. The low ceiling and tin roof made it almost unbearable in summer. Pouring water on the roof provided temporary relief and placing a bamboo screen on the window blocked the worst of the sun's rays. None of this bothered Iida. As an entrepreneur he was achievement-oriented. His surroundings were of little consequence so long as he had the tools to get the job done. Iida often spent the night circulating among his clients' premises, supervising the guards and advising them as to how they could best tackle the problems they encountered. Toda, meanwhile, would remain at the office to fill in as dispatcher. Despite the poor working conditions spirits remained high among the employees.

The people in the surrounding area, including the police, were curious about the activities of the uniformed men who would come and go from the building. From time to time Iida and his staff would be queried as to their line of work, particularly when one or more of the men was seen repairing the secondhand Datsun (Nissan) Bluebird automobile with the unfamiliar logo on its doors that served as the patrol car. Motorcycle guards conducting late night checks were frequently stopped by the police and questioned as to their activities. Although they carried identification, providing a satisfactory explanation was sometimes difficult since the police had seldom heard of security guards and knew nothing of their function. As Asado recalls:

> Before my time some of the patrols were made by bicycle but when I joined the company the guards were using small Honda motorcycles. We had trouble with their engines when it rained and in the winter. Often they would not start and we had to make our rounds by taxi, an expensive alternative. We didn't have radio communication with the dispatch centre in those days and were dependent on pay phones in order to make our reports. Mr Toda worked in the office during the day and slept there at night so that he could serve as dispatcher. Mr Iida concentrated on sales during the day and at night he toured the clients' premises to give encouragement to the guards and ensure that they were alright.

About the same time as the move, Nihon Keibi Hosho obtained a contract for round-the-clock surveillance of Kokusai Tenjijo, the international exhibition grounds in the Harumi district of Tokyo. As there was no on-site accommodation for the security guards, Iida provided a tent in which the guards could stay in their off hours. Tadao Hayashi was hired in April 1963 to work at this site:

> Today there are some impressive buildings on the exhibition grounds but in the early sixties the site was mostly open field. Automobile shows took place out in the open. There was no place for the guards to take shelter so we pitched a tent in which

to rest at night. The site, adjacent to Tokyo Bay, could be extremely cold on rainy, windy days. Often, the charcoal heater wasn't enough to keep us comfortable so we would use our nightsticks as swords and practise kendo swings to keep warm.

Conditions were far from desirable but Iida and his employees were content in the knowledge that the security business was gaining recognition in Japan.

With the business growing, in order to maintain a clear financial picture useful also for the development of future strategy, Iida felt it would be advantageous to establish the practice of an annual external audit. There were many external resources from which to choose. Iida examined several domestic auditing firms but eventually decided on the Japanese arm of an international firm, Price Waterhouse. The company had begun to operate in Japan the year Iida founded Nihon Keibi Hosho. In 1962 Price Waterhouse merged with the auditing firm of Lowebigham and Thomson, operating in Japan since 1949. Iida explains:

I have been using Price Waterhouse since the second year of the company's establishment. They are more expensive than many Japanese firms, but I feel it is money well spent. With this firm there is no compromise. The audit is severe and clearly shows where the company stands. Not only with auditing but for any work that requires analytical thinking, I prefer to use the best people available.

TOKYO OLYMPICS

Japan had been selected to host the 1940 Olympic Games but World War II intervened. For twelve years, from 1936 until 1948, no Games were held. A decade and a half later, in the midst of its postwar recovery, Japan was ready to show the world its new face. The country was chosen to host the Eighteenth Olympiad to begin on 10 October 1964.

In preparation for an event of this magnitude many facilities had to be built. Ensuring the security of these sites throughout their construction as well as while the Games were in progress was a major responsibility confronting the Tokyo Olympic Organizing Committee. The National Police Agency which included the Tokyo Metropolitan Police Force, the National Self-Defence Force, and the National Fire Defence Agency would be enlisted for this and many other purposes including the running of the Olympic flame from Okinawa to Tokyo but the manpower from these sources would not be adequate to provide maximum security.

Of particular concern was the Olympic Village where the athletes would be housed throughout the Games. The Village was to be located on a 350,000 square metre tract of land in central Tokyo taken over by the Americans during the occupation. The area was named Yoyogi Washington Heights and four hundred houses were erected to accommodate American military families. On the occasion of the Olympics the area was returned to the Japanese government and designated for use as the Olympic Village. When completed it would accommodate over 6,000 people.

Juzaburo Okabe was responsible for overseeing the construction and security of various Olympic facilities:

> One of my biggest worries was how we were going to provide adequate security for the Olympic Village not only when the Games were in progress but also during the preparation of the site. I had no idea where we would find the necessary manpower. Then I heard from the Tokyo Metropolitan Police Agency and from the Tokyo Metropolitan Fire Board that there was in the city a company specializing in security services.
>
> I contacted Nihon Keibi Hosho and was surprised by and impressed with the thoroughness of the company in cost estimation, its personnel management and the security operations. All my problems relating to the provision of security were suddenly resolved. The Olympic Office had submitted a written pledge to the Japanese government promising that the Olympic facilities would be properly secured prior to and during the Games. I was confident that if the security of the Olympic Village and the competitive sites was turned over to Nihon Keibi Hosho the Olympic Office could live up to its pledge.

In November 1963 Nihon Keibi Hosho was formally asked to assist in the provision of security at the Olympic Games. Iida was excited and felt that this was the contract that would link irrevocably Nihon Keibi Hosho with the security business in Japan. Nevertheless, he had reservations about accepting it. A large number of men would have to be hired, at considerable expense in terms of equipment and training. Their employment would be for less than a year and when the games were over they would have to be let go – an undesirable situation in a country where life-time employment was a common practice.

Weighing the pros and cons, Iida decided to go ahead with the

3 Nihon Keibi Hosho's guards arriving at the site of the Tokyo Olympic Village, 1963.

negotiations. While the two-year contract had to be waived, he was adamant about the three-month payment-in-advance. Initially the Tokyo Olympic Committee was unwilling to comply with this condition but they had little choice. Adequate security was necessary and no other company could provide it.

As soon as the contract was signed Nihon Keibi Hosho hired forty-five additional men. The security guards already on staff played an important role in training the new recruits in the practical aspects of guard duty. At noon on 10 December 1963 fifty Nihon Keibi Hosho guards led by their corporate flag and carrying a replica of the Olympic flag entered the central compound of the future Olympic Village. The Stars and Stripes was lowered from the flag pole and the Olympic flag hoisted in its place. Several months later, with the arrival of the athletes for the games, Nihon Keibi Hosho's guards would be augmented by 136 police officers.[11]

While no major incidents were encountered in the process of guarding the Olympic Village prior to the opening of the Games, the work was not uneventful. Curiosity seekers and potential thieves would illegally enter the village to see how the foreigners had lived and in hopes of finding valuables. When apprehended some of these intruders turned out to have criminal records. Young lovers were also among the illegal entrants, looking for privacy and space, rare commodities in urban Japan. As the area had been under American control for sometime the surrounding neighbourhood was devoid of the traditional police boxes essential to the smooth functioning of Japanese urban communities on a daily basis. Nihon Keibi Hosho guards stepped in to fill the void, stopping street fights, escorting pregnant women to the hospital, and protecting women against molesters.

The salaries and other expenses of the expanded work force severely strained the financial resources of the company. Twice prior to the Olympics it appeared that the company would not be able to meet its payroll commitments on time but in each instance new contracts with large corporations, and therefore additional funds, were secured just in time.

Iida worked hard to negotiate as many contracts as possible in order to provide the maximum employment opportunity for the Olympic guards once the Games concluded. The fact that Nihon Keibi Hosho had been entrusted with the provision of security for such an important international event as the Olympic Games and the recognition this security received in the mass media bolstered the company's credibility. The Olympics were a success. Ninety-four countries were represented with 6,550 athletes participating in the games. For the forty days from the opening to the closing of the Olympic Village no serious incidents occurred.

Nihon Keibi Hosho became widely known among civilians and policemen alike. Fumitaka Ito, a reporter with a police newspaper, *Hoan Keisatsu Shimbun*, ascertained from the many police officers with whom he came in contact that there was widespread acceptance in the police force of the emergence of a private security corporation and high expectations for the future contribution of the company to security. The police welcomed the 'competition' from the private sector and looked forward to future cooperation.

High ranking police officers frequently queried Ito regarding the type of training given to the Nihon Keibi Hosho guards. Ito had been observing and writing about Nihon Keibi Hosho almost from its inception. He felt that the officers were of the opinion that the employee education and training would be closely linked to the quality and growth of the company. Following the Olympics, security firms sprang up throughout Japan. Employee education became their key to success.

The year of the Tokyo Olympics saw the completion of the Tokaido Bullet Train, allowing Japanese greater mobility. Improved highway networks facilitated automobile travel and the number of cars per capita steadily increased. These factors may have contributed to the marked expansion of the geographic area in which given criminal activity occurred. This expansion was particularly noticeable in the mid-sixties.[12]

GROWTH

In the booming economy of the mid-sixties, there was a growing awareness on the part of Japanese corporations of the desirability of externally provided contract security instead of in-house security. Before the Games ended Nihon Keibi Hosho had more contracts than Iida had thought possible. Just before the opening of the Games, Isetan, a major Tokyo department store, contacted Iida inquiring about the type of service Nihon Keibi Hosho could provide and subsequently signed a two-year contract. Shortly after, the Imperial Hotel in Tokyo followed suit. This was followed by a contract with Tokyo Gas Company.

Another client was the Tokyo branch of the Chuo Keiba Horse Racing Association. Legalized betting was on the rise in Japan particularly on horse races, bicycle races, motorcycle races, and speedboat races. As the crowds increased in size they became more rowdy and destructive at these events, often damaging facilities and sometimes setting fires. In 1962, the year Nihon Keibi Hosho was established, there were ten such incidents and a year later, twenty.[13] The police requested that the ministries responsible for the various sporting activities (Agriculture and Forestry for horse racing; International Trade and Industry for bicycle, motorcycle, and automobile racing; and Transportation for speed boat racing) should work with their respective racing operators to improve race management and enhance crowd control through facility upgrading.[14]

The police also advised increased on-site security. Especially at horse racing tracks, not only was it necessary to maintain order among the crowd, it was also essential to guard against fire in the stables, the human sleeping rooms/quarters, the feed rooms, storage rooms, blacksmith's shop, kitchen, mechanical equipment room, and the toilets. Chuo Keiba Horse Racing Association was the first of many such organizations to hire external security services as racetrack betting increased in popularity. In 1964, 3.5 million people attended horse races throughout Japan. Ten years later the annual attendance had risen to 14.6 million.[15] With the larger crowds security measures became essential.

Many other contracts followed for Nihon Keibi Hosho. The demand for security had become so great that it was unnecessary to lay-off any of the Olympic guards at the conclusion of the Games. Indeed, one hundred additional guards were quickly hired to meet the firm's increased personnel needs. By the end of 1964, less than three years after the company was established, the number of employees had risen to three hundred. Makoto Iida's dream of breaking new ground in the business world was coming true. He had taken the initial steps along the path he was forging for the security business in Japan.

From the knowledge gleaned during the first three years of operation Iida and Toda developed an understanding of the differences between in-house security and the services provided by a professional security firm. They came to realize that there were four essential operational components to the provision of comprehensive security. Firstly, there was the identification of potential and actual problem areas at the site: unlocked access points, fire hazards, etc. Next was the correction of these problems. This was followed by the determination of propensities and trends made possible through the collection and analysis of statistical data. Finally there was the establishment of measures and strategies for incident prevention.

While the first two components fell primarily within the domain of on-site guards, the latter two were the responsibility of management. Iida and Toda came to understand that the provision of a system for economical, effective, and easily actualized data collection and analysis, the communication of the findings to the guards, and the formulation of the best preventive measures in cooperation with clients was a key strength upon which a professional security firm could capitalize. Through involvement in different cases information could be synthesized and integrated so that a more comprehensive approach became possible. This holistic approach was often lacking in in-house security and gave Nihon Keibi Hosho a distinctive stance.

One of the internal problems confronting Iida at this time was the amalgamation of Nihon Keibi Hosho's two labour unions. The first union had been formed in 1963 under the leadership of Genzo Seki, a forceful individual but one who listened to reason. When Seki and his supporters postulated the idea of a labour union Iida offered his support: 'I felt that the formation of a labour union was necessary for the healthy development of the corporation. It would prevent me from acting unilaterally.'

With severely limited capital, salaries were kept to a minimum in order to create profit which could then be used as operational capital for corporate advancement. Iida ensured that his employees were aware of the company's tight financial position. Sympathetic to the situation, the union did not try to force the issue through strike action.

However, not all employees were happy with what was perceived by some as the union's lukewarm approach and disgruntled workers, many of whom had been hired to provide the personnel necessary for the company to fulfil its Olympic contract, formed a second, more militant union. With the Japanese economy taking off and substantial profits being made by established corporations, workers throughout the country were beginning to realize that their contribution to the nation's economic revival was not being adequately rewarded. One recourse was to strike. Because Nihon Keibi Hosho was new and fragile and because the provision of its service depended solely upon its workforce, Iida feared that a strike would finish the corporation.

The emergence of a second union within a Japanese corporation is relatively common and frequently leads to the disintegration of a strong union position, giving management the upper hand. This stems from the inherent character of the labour unions themselves. Japanese unions are corporate labour unions. New employees are automatically members and anyone resisting membership is considered to be an extremist. This often results in a lack of interest and commitment among union members. Union and management are closely linked and it is not unusual for a corporation to provide the physical and monetary resources necessary to support the union in its negotiations with management.

Some argue that labour unions in the western sense do not exist in Japan and that the position of union boss is the equivalent to an executive position responsible for labour relations.[16] When labour/ management confrontation intensifies, a second labour union will often emerge either as result of management manipulation or the dissatisfaction of the rank and file with union leadership. Sympathetic to management's position, this new labour union will protest the strong stance taken by the original union. The second union will quickly gain support, bringing about the dissolution of the first union.

Nihon Keibi Hosho's position was the reverse. The second union was strongly anti-management. Iida realized that the labour union was the vehicle through which labour negotiations could and should be carried out but the existence of two unions made this difficult. For several months Iida devoted his energies to bringing about their reconciliation. During the course of this process Genzo Seki was hit by a truck and killed while on duty in the Olympic Village. This may have facilitated the coming together of the two unions as, in the absence of a leader, members of the original union were less opposed to reunifica-

tion with their fellow workers and at the same time supporters of the new union were sympathetic to the loss suffered by members of the first union.

May is the traditional time for labour negotiations to take place in Japan and in May 1964 Iida was negotiating the yearly bonus with representatives of the two unions who were threatening strike action. During the negotiations Iida received a telephone call from his mother informing him that his wife had just given birth to a baby girl. Hanging up the phone, he shared the news with the union members. They were impressed and touched that he had devoted himself to the negotiations without mentioning the impending birth.

The issue of the bonus was quickly resolved to the satisfaction of all sides and shortly afterwards the two unions were united into one. The birth of Iida's second child also helped to ease the rift between Iida and his father. Iida and his wife named her Mami, derived by combining the first sounds of his father's and mother's given names.

Nineteen sixty-four was also the year that Toda married. His bride was Akiko Koikegami whom he met while working at the travel agency. When he proposed to her he reminded her that his future was uncertain and that he was a Catholic. Undeterred, she accepted his proposal. She studied Catholicism and was given the name Maria at the time of her conversion. Soon after the wedding Toda put his bride and all their possessions into a small pickup truck and left Tokyo for Osaka to establish Nihon Keibi Hosho's first branch. Reminiscing about that time Mrs Toda says,

> There was no room for our futon [bedding] in the small truck so we shipped it separately. It was evening when we arrived in Osaka and discovered that our futon had been delayed. We were tired after the journey so we went to the local shop and bought the necessary bedding. As soon as we got it home and spread out on the floor we fell asleep.

Once the nation's economic heartland, today Osaka is Japan's second largest commercial centre. Such corporate giants as Sanyo, Sharp, Sumitomo, Suntory, and Matsushita (National) are headquartered there. Osaka is home to the world's third largest stock market as well as some of its major banks. When it opened on 1 October 1964, the Osaka office had only four employees: Toda, a female clerk, an experienced guard from Tokyo, and a male 'jack-of-all-trades'.

Initial acceptance of the service Nihon Keibi Hosho had to offer was somewhat more rapid here than it had been in Tokyo. Not only had the company's reputation preceded it, but also Osaka lived up to its reputation for being more open to new ideas than Tokyo. Financial institutions and high rent apartments were among the first customers and additional guards were hired as required. Says Toda, recalling the period:

> We were newly married but I felt that building the business had to come before everything else. I placed the telephone on the floor beside the futon on which we slept and our room became the company communications centre each night. The night guards would phone in their reports regularly and I would respond and update their orders.

Mrs Toda adds:

> I understood that my husband had to be on-call twenty-four hours a day. This was a difficult time in the life of the company as it was not yet firmly established. Its success depended on the unfaltering dedication of all those involved with it.

Toda remained in Osaka for six months and when he felt that the branch was on a secure footing he returned to Tokyo. Following a similar course of action, a branch was opened in Nagoya in May 1965 and the following year four more offices were opened in Fukuoka, Kobe, Sapporo, and Yokohama. This was the starting point for the development of a nationwide service.

Not only was Iida attempting to instruct clients and potential clients in the desirability of adequate security measures, he was at the same time learning from his clients. One such client was Tetsuzo Inumaru, president of the Imperial Hotel.

Born in a rural village in Ishikawa prefecture in 1887, Inumaru studied commerce at what today is known as Hitotsubashi University in Tokyo. Upon graduation in 1910 he went abroad to study hotel management in Manchuria, Shanghai, London, and New York. When the decision was made in 1919 to redesign the Imperial Hotel, established originally in 1890 as Japan's first high class western style hotel to cater to foreign dignitaries, Inumaru was invited to become the assistant general manager of the project. Frank Lloyd Wright was to be the architect but as this was his first hotel someone with a thorough understanding of hotel operations and fluent English was required to work closely with him. Inumaru returned from the Waldorf Astoria Hotel in New York to assume this position. When the new hotel opened in 1923 he became its general manager and twenty-two years later assumed the presidency.[17]

During his career Inumaru served as chairman of the Japan Hotel

Association and contributed to the establishment of a number of fine
hotels in other major cities in Japan through the provision of
management know-how and personnel. His long experience in hotel
administration had rendered him well-versed in every aspect of facility
management. Recalling his association with Inumaru, Iida says:

> One evening soon after we began providing security for the
> Imperial Hotel in October 1964, I received a telephone call from
> Mr Inumaru. Along with my entire staff I was excited but at the
> same time apprehensive about providing security for Japan's most
> prestigious hotel and our first client of this type. When Mr
> Inumaru asked me to come to the hotel I rushed over
> immediately, fearing that there had been some error on the
> company's part. My concerns were soon put to rest when Mr
> Inumaru greeted me with the utmost friendliness and offered to
> instruct me in the details of hotel security. He took me on a tour
> of the exterior and interior of the hotel pointing out the dis-
> tinctiveness of hotel security. By the time we finished it was after
> ten o'clock. Mr Inumaru was in his late seventies at the time. I was
> impressed with his commitment to the safety of the hotel guests
> and his willingness to share his specialized knowledge of hotel
> security.

Hotel security differs markedly from industrial security. Large hotels
are, in many ways, like miniature cities and have all the attendant
security problems that a city might have. Due to the nature of their
business hotels implicitly invite all to enter and utilize the lodging
facilities. To reject anyone who is legitimately looking for a place to
stay is difficult. Because of the diversity of security needs within a
hotel both contract and in-house security can be effective. At the
Imperial Hotel Nihon Keibi Hosho guards were used, among other
purposes, to guard the employee entrances and shipping and receiving
areas and to patrol the hallways, freeing the hotel staff to handle the
more delicate areas relating to problem guests. The contract with the
Imperial Hotel paved the way for a number of other hotel contracts.

This particular working environment presented special problems,
the solutions to which provided both Iida and his employees with a
wealth of experience. Yoshinao Asado who was assigned to a Tokyo
hotel in 1965 comments:

> I learned many things from providing security at a hotel. At the
> hotel where I worked the guests were almost always American
> soldiers on leave from Vietnam. Often they'd drink throughout

the night and create a disturbance. Others would go to sleep and have nightmares about being attacked by Vietcong. Still dreaming they'd get out of bed and try to escape from the hotel's emergency exit in their underwear. Without adequate English it was very difficult to effectively deal with these problems. I felt strongly that diverse experience and knowledge were essential to perform well as a security guard.

English language fluency was required especially by guards working in the large city hotels. Iida recognized this and English conversation became part of the training curriculum for hotel guards. A few months later it was integrated into the training for guards assigned to security duty at international airports and ports.

Not only has Iida's work been made easier through clients like Inumaru of the Imperial Hotel, his private life has also been enriched. One early client to whom he is indebted is the late Noboru Goto president of Tokyu, one of Japan's leading railway companies and the nucleus of the Tokyu Group of some three hundred firms. From Goto he learned how to involve himself wholeheartedly in leisure activities when the time was right. The two enjoyed golfing together and in the late seventies Goto introduced Iida to troll fishing. The sport has since become a summertime passion. Goto's death in 1989 was a great loss for Iida.

'FAME'

An unexpected spin-off of Nihon Keibi Hosho's Olympic involvement and subsequent mass media exposure was a request from TBS (Tokyo Broadcasting Station) television to dramatize the work of the company in a weekly television programme. Cognizant of television's influential role in image formulation among the general public, Iida consented on the understanding that the programme would accurately reflect security work and that he could have a say in its content. He insisted on a number of conditions. No offensive language was to be used; women were not to be degraded; and the consumption of liquor was not to be portrayed. He also specified that scenes were to be shot in modern settings – for example in newly built western style hotels rather than in traditional Japanese *ryokan* – in order to create the image of a thriving, up-beat service.

The broadcast corporation suggested two possible titles for the programme: *Tokyo Yojinbo* (Tokyo Bodyguards) or *Tokyo Yakeisha* (Tokyo Nightwatch Company). Iida objected to both of them.

4 Iida (far right) and Toda (back row left) with the cast of the TV drama 'The Guardman', 1965.

Neither, he felt, portrayed accurately the image of his company's activity. He consulted with Toda and together they hit upon the westernized name, The Guardman, pronounced in 'Japanese' *Za Gardoman*. Iida agreed to the use of Nihon Keibi Hosho's stylized owl to mark the patrol cars in the drama, as it did in real life.

Broadcast of the drama began in April 1965, at a time when 95 per cent of Japanese households had acquired black and white television. The cast of familiar actors including Ken Utsui, Jun Fujimaki, Yusuke Kawazu, Yoshio Inaba, Isao Kuraishi, and Shizo Chujo coupled with the innovative script made the programme one of the most popular. Dramatization of their work was a source of pride among Nihon Keibi Hosho's employees and their families and bolstered the image of security guard duty. While the exposure helped familiarize the public with the company, the image presented of a labour intensive, high risk security firm had an adverse effect on loan procurement and employee recruitment. The dramatization coupled with Nihon Keibi Hosho's business success also triggered the establishment of a rash of security businesses throughout Japan.

Sogo Keibi Hosho, Japan's second security company and today SECOM's number one rival, was the first to emerge. Its president was Jun Murai. Born in 1909, Murai graduated from the Faculty of Law of

Tokyo University in 1935. In 1946, he was appointed secretary to Prime Minister Shigeru Yoshida (1878–1967)[18] and subsequently served as director of research in the Prime Minister's office. In 1958, he assumed the position of Commander-in-Chief of police forces in the Kyushu district. Four years later he was appointed to the administrative office of the Tokyo Olympic Organizing Committee.

In December 1964, Murai accepted an invitation to have lunch with the then retired Yoshida, thinking that the former Prime Minister wished to show his appreciation for the work he had done for the Tokyo Olympics. Instead Yoshida asked Murai what he planned to do now that the Olympics were over. Murai had been wondering about this himself. Many suggestions had been put forth ranging from becoming an executive with a public corporation to establishing a private security corporation. To Yoshida's question regarding the existence of security firms in Japan, Murai replied:

> Two or three years ago a Swedish security company successfully established itself in Japan. I have also heard that such first rate companies as Securico of England and Brinks of the United States have plans to enter the Japanese market.[19]

Yoshida's response was immediate:

> If this happens security nationwide will soon be controlled by foreign capital. Forget about accepting a position with a public corporation. You should create a wholly-owned Japanese security corporation. There is no room for deliberation.[20]

During the drive home Murai pondered Yoshida's advice. How could he, a civil servant, succeed in the private sector? Perhaps, he thought, he could learn from the determination and spirit of the military lords in the age of civil strife during the feudal period. Theirs were life and death decisions. A mistake could result in the death of thousands, themselves included. How much greater was their responsibility than that of a contemporary corporate manager. A mistake on the part of the latter could be alleviated by a letter of resignation or the declaration of bankruptcy. Before he had reached his destination he had determined to adopt the spirit of the military lords and establish a security corporation.[21]

In preparation for the new venture Murai reviewed mentally the concerns to which a feudal lord would attend preparatory to battle. The following seven points came to mind:

1 Clear representation of the cause for raising an army.

5 Nihon Keibi Hosho's 'armoured car', 1966.

2 Installation of a respected general.
3 Recruitment of a clever tactician capable of the development of foolproof strategy and tactics.
4 Procurement of campaign funds and provision of adequate weapons, ammunition, and food.
5 Cultivation of as many allies as possible.
6 Recruitment of competent leaders and strong soldiers to create a fighting force of maximum strength.
7 Examination and analysis of various situations particularly the conduct of the enemy as the battle approached.[22]

Murai persuaded Daigoro Yasukawa (1886–1976)[23], former chairman of the Tokyo Olympic Organizing Committee, to become the Chairman of the Board of the proposed company and procured the necessary bank loans. A corporate symbol was chosen: an eagle representing strength standing on a nightstick, its straightness symbolizing accuracy, backed by the sun (sincerity) and partially surrounded by laurel leaves (speed).[24]

On 16 July 1965, with twenty-eight employees, Sogo Keibi Hosho came into existence. Today the company's employees number 21,000.[25] While both Nihon Keibi Hosho and Sogo Keibi Hosho moved in the direction of security systems integrating people and technology, the emphasis at Sogo Keibi Hosho has been primarily on human-based security service. The company's human resource development programme stresses employees' spiritual growth, commitment to the organization, and indebtedness to the customer.

Several more security firms emerged over the next few years, all of them smaller than Sogo Keibi Hosho. A few began with only one or two employees and a telephone. By 1989, security firms in Japan numbered 4,900, employed 219,000 people and their sales totalled 901 billion yen.[26]

While Iida's company has grown significantly over the years, he has been acutely aware of the need to differentiate between 'growth' and 'putting on fat'. Growth is an economic rather than a physical term, to be measured in terms of contribution to society, productivity of resources, and profitability. Iida is particularly careful not to consider increase in employee numbers as a growth indicator. From the company's inception, the challenge has been to provide services in a way that would create the optimum balance between risk and return on resource input.

Competition may be a double-edged sword for entrepreneurship. Insufficient competition makes growth and innovation easier but at the same time less essential for survival. On the other hand, excessive competition requires more innovation and expansion if the entrepreneur is to maintain his niche but at the same time it makes these activities riskier because of the greater likelihood of losing one's market share.[27] Competition, which originated with the founding of Sogo Keibi Hosho and has continued to grow through the ensuing decades, has, by and large, been a positive influence on corporate development. Iida has been challenged to search for new and better ways of providing security services in order to maintain his position of prominence within the industry.

5 Technology-based innovation

REMOTE ALARM SYSTEM

Iida's early business success can be attributed to his ability to perceive a familiar situation in a new light and his subsequent decision to enter into computer integrated remote alarm security at a time when this was not yet a reality in the technological or even the conceptual sense. It was an innovative idea stemming from Iida's ability to question traditional ways of doing things and his open thinking – sometimes perceived by those who 'knew better' as naive credulity – towards new notions and methods. Equally crucial to innovative success was his sense of timing.

The objective of the security industry is to provide maximum security at minimum cost. Despite not having a technical blueprint, Iida and Toda had contemplated for sometime a marriage of alarm and communication systems built on electronics technology in order to permit off-site observation of guarded premises. Their goal was to create an integrated system where electronics and personnel would work together in the customer's best interest. The electronic system would extend the range of human sensitivities enormously, allowing a single guard to be responsible for the security of a much larger area than was possible when he had to rely solely on his five senses to tell him whether or not a particular location was secure. Initially, the lack of recognition accorded the security business coupled with Nihon Keibi Hosho's tenuous financial situation prevented exploration of this apparently radical idea.

Eventual entry into the field of remote security was initiated by the company's astonishing Olympic-triggered growth. Iida was confronted with the challenge of maintaining a uniform quality of service, difficult when service was dependent almost entirely on the human factor. One problem was morale. Guard duties were monotonous and routine, requiring patience in their execution. Paradoxically, the lack of dramatic occurrences was the 'ideal' state but at the same time this 'ideal' state fostered boredom and a decline in morale.

From the client's point of view, it was expected that a professional guard would be motivated and have a high morale. However, declining morale permeated routine guard duties and, while it was not necessarily reflected in a lowering of service, it often adversely affected congenial client/guard relationships. This problem was compounded by the fact that in Japan, even when queried, a customer would never talk directly about the shortcomings of a service, especially when the dissatisfaction might stem from displeasure with the attitude of the service personnel.

Iida and Toda realized that it is essential for the management of a service industry to be sensitive to customer needs and to recognize the degree to which customers are dissatisfied. In the service industry, this requires frequent visits to the site to observe the employees' work patterns and behaviour and the clients' response to and interaction with them. The manager must have the necessary skills to identify problems and then must take steps to rectify them through ongoing communication and training. The larger Nihon Keibi Hosho's guard force became, the greater would be the potential for client dissatisfaction.

Another problem was associated with the fact that professional security was a new occupation and therefore was without a formally or informally established code of conduct. In addition, employees had come from other companies in both the private and public sectors, each with his acquired values and behaviourial patterns from his previous employer. It fell to Iida and Toda to teach their employees professional ethics and orient them to Nihon Keibi Hosho's corporate philosophy. Yet no amount of education could completely eliminate problems stemming from the 'human factor'.

Iida had begun to doubt the efficiency of the human-based security system. Working to improve the established system was not good enough. He wanted to replace it with something superior. It was out of his ability to entertain doubts expressed towards the existing service that a new system emerged.[1]

Iida approached a number of manufacturers seeking one that would develop a remote alarm system. As he did not have a detailed understanding of the hardware or technology his vision was not constricted by the technological limitations of the time. He turned his lack of concrete knowledge to advantage, focusing on the ideal integrated security system based on technological potential rather than on technological limitation.

The manufacturers approached did not jump at the opportunity Iida presented to them. The idea was ahead of its time. Even direct emergency communication in the form of a three digit telephone number was not widely available in Japan. Introduced in Tokyo about 1955, it would not be available throughout the nation until the late sixties.[2]

Shiba Electric Corporation, later taken over by Hitachi Electron, was the only manufacturer to express interest in Iida's project. At that time Shiba Electric was an independent company unaffiliated to any corporate grouping. With 1,000 employees it was engaged in the manufacture of broadcasting equipment. Sadaharu Chiba of Shiba Electric was responsible for the development of the prototype for the remote alarm system. He was in charge of system design and adjustment for television and radio broadcast equipment. A major project of his was the development of a remote control system that would permit the control of mountain-top broadcasting equipment from the broadcast studio using signals transmitted along the private circuits of the government-owned telecommunication company, Nippon Denshin Denwa Kosha (Nippon Telegraph and Telephone Public Corporation). As Chiba recalls:

> Because of the perceived similarities between the remote control system I was working on and the remote alarm system Iida wished to develop, we entered into research on a remote surveillance system that would detect illegal entry and fire.
>
> Because this was the first of its kind in Japan, I couldn't visualize the final product. Iida had an image of it in his head and I was responsible for making it a reality. My method of work altered. Instead of sitting at a desk, I walked around outside wrestling with Iida's concept until I was able to envisage what I thought would be an acceptable final product. We submitted a cost analysis and specifications for what amounted to a miniature version of the remote control system we had developed for use in broadcasting.
>
> Iida rejected it on the grounds that its retail cost would be too high and its surveillance capabilities were minimal. I had been accustomed to dealing with the government and other large corporate clients to whom cost is of much less importance than it is to individual consumers. Iida was still convinced that we could make what he wanted. Following many consultations, eight months after he first approached Shiba Electric, we had developed a remote alarm system that finally met his require-

ments. It included such unlikely components as a motorcycle key.

From the outset, when Iida first made me aware of what he was seeking, I felt that he already had a larger plan. Even then, I think, his vision took him beyond machine security to the concept of a communication network service.

In 1971 Chiba left Shiba Electric and joined Nihon Keibi Hosho's research group. Today he is the assistant manager of the company's Technical Centre.

The pattern of development of this first alarm system typified the development of all future innovative products and services. Iida plans what he wishes to accomplish and selects the appropriate personnel in order to see the project through to completion:

Consider the extensive integration of computers into SECOM's systems. Even today I know almost nothing about the technical side of computers. When I look to the future and see a new opportunity to utilize the potential capability of the computer, I put together a team to actualize the computer integrated system. I give them only the concept I envision. The team's mandate is to create the reality.

Often the initial reaction is bewilderment about how to proceed. My role, I feel, is to read the future and to conceptualize future business opportunities. The role of my staff is to realize these opportunities. I ask the chosen employees if they can do what I have requested. If someone lacks the necessary confidence I will find another to carry out the task at hand. When they agree that it can be done I entrust them with the project completely and move on to do my own job of conceptualization of innovative ventures.

The integration of electronics technology with the security guard service, Iida thought, would not only improve the quality of the service but would minimize the personnel problems which might increase in the future if the company continued without alteration. The conceptualization and realization of a security system integrating people and technology became the catalyst for subsequent dramatic changes in the way in which Nihon Keibi Hosho's service would be provided. It was the beginning of a transformation from the 'old' to a not clearly defined 'new' corporate activity.

Iida christened the remote alarm system, intended for institutional security, the SP (Security Patrol) Alarm System. Its communication circuits were private lines leased from Nippon Denshin Denwa Kosha. He debated whether to sell or lease it to clients and decided in favour of the latter as it would afford him greater control over the maintenance and upgrading of the system. Such control was essential if Nihon Keibi Hosho was to provide the highest quality service at what, in the long run, would be the lowest possible cost. If the system was sold outright to clients it would be difficult to ensure the replacement of outdated components. As security systems became more sophisticated, so too would the techniques employed by criminals and this in turn would spur further development of the system. Variations in the adoption of equipment among customers would make integrated central security control difficult.

As well, leasing the equipment on two-year contracts, while it would garnish less immediate income than outright sales, would provide a more predictable financial base. Assuming that few clients would decide against contract renewal, the tighter profit control would make investment in new security business ventures less risky. Having decided in favour of equipment rental Iida looked closely at Xerox and IBM, two giants in the rental field, and decided on his financial course of action.

Iida and Toda, accompanied by their wives, first visited Sweden and then travelled to New York to attend a meeting of the Ligue Internationale des Sociétés de Surveillance. It would be another two years before Nihon Keibi Hosho would be in existence long enough to fulfil the company age membership requirement. For this meeting Iida's status was that of observer but he was given the opportunity to share with senior members of the security business his plans for the SP Alarm System. Confident of his scheme, he shared with them the concept of people-technology integration upon which the system was based and his intention of leasing rather than selling the system's hardware.

The response was not what he had anticipated. Amid much laughter the fifteen members present were quick to point out the system's perceived shortcomings. They cited the high risk involved in areas such as hardware malfunction and various issues associated with compensation. Iida was naive, they felt, to think he could insist on leasing the equipment. If clients wanted to purchase it outright he would have to comply.

Afterwards Iida reflected on the response of his colleagues. Maybe he was wrong. Maybe he had deluded himself into believing in the success of a scheme that was impossible to realize. 'I gave the situation a great deal of thought and concluded that Nihon Keibi Hosho was not, first and foremost, selling equipment. The company's mission was marketing safety.'

Iida's intention of entering the field of machine security derived not only from his desire to capitalize on technological potential but also from his perspective towards the place of human beings within the organization. He wanted, whenever possible, to eliminate from the routine of his employees those mundane, boring, repetitive, and dangerous tasks which could be handled by machine. 'Employees freed from repetitive work can be engaged in decision-making activities for challenging projects. Placing people in jobs which can be done by machine is not only a waste of resources, it is a desecration of human beings.' Today, much repetitive and dangerous work in production processes is given over to industrial robots[3] but the idea of a machine doing a person's job was very new in the mid-sixties, especially in the service sector.

Iida's faith in his idea remained intact. On his return to Japan, the next task was to determine the best route by which to introduce the electronic system into the marketplace. Iida's choice was to tackle first the most resistant institutions – the banks. If he could penetrate this market other clients would be forthcoming. About 1962 the National Police Agency began providing banks and other financial institutions with security inspections and security guidance services although these were not systematically instituted.[4]

Recognizing that their adversaries are seldom amateurs but are professional burglars interested in big jobs with high risk/yield ratios, banks had developed tight internal security systems of their own. At first they could not see the need for change, especially change that would require the intrusion of 'outsiders'. Iida met with comments such as, 'We cannot entrust our security to a machine like "an electric mouse trap", or "Machines". We feel more secure with guards on the premises round the clock.' Nevertheless, as a result of Iida's determination to find a user and his fervent belief in the newly developed system, six weeks after the prototype was completed in May 1966, Iida secured a contract. The first SP Alarm System was installed at the Ikebukuro Branch of the Mitsubishi Bank in Tokyo.

Kiyoshi Okada had joined Nihon Keibi Hosho in 1964 as an on-site guard and was transferred to the company's main office in January of 1966 to work in the dispatch centre. At that time the company had twenty contracts for on-site security guards and another fifty contracts for periodic security checks. Guards were dispatched in patrol cars, on motorcycles, and on bicycles. The latter two had the advantage of easy manoeuvrability through congested areas and required no licence to operate. They were the preferred mode of transportation if the client was located close to the dispatch centre. Okada recollects:

> It was 20 June 1966. Unannounced, three electricians came into the office and installed a strange machine. Iida met with me and three colleagues. I was told it was my job to constantly monitor the machine. If a red light went on it would indicate that something was amiss at our client's premises. In this case the client was the Ikebukuro Branch of the Mitsubishi Bank. If the machine indicated a problem my colleagues were to be dispatched to the site immediately. This was a trial run and we were jubilant when the signal came through. The system really worked!

More contracts followed. Shinichiro Osada, a friend of Iida's from junior high school who had gone to work for the Nippon Television Broadcasting Corporation after graduation from Keio University, joined the company's sales department in 1966. After ten years in programme production Osada felt that the work was no longer challenging. He favoured an increase in the broadcasting corporation's documentary production. Management decided otherwise. The emphasis would be on entertainment. Aware of his dissatisfaction Iida suggested that he might be interested in working in public relations and sales at Nihon Keibi Hosho.

Iida, Toda, and Osada concentrated their efforts on selling the SP Alarm System. That year thirteen contracts were signed. The next year the contracts for the SP Alarm System totalled fifty-nine and the following year 165. By 1969 they numbered just over 500 and by 1970 approximately 1,400 contracts had been negotiated since the introduction of the system. Throughout the ensuing decades the SP Alarm System has continued to grow in popularity and is now widely used in financial institutions, factories, and government buildings, even the Prime Minister's official residence.

With the business growing, in 1969 Nihon Keibi Hosho purchased its first computer, a FACOM 230-25, in order to streamline office routine. Made by Fujitsu, it was one of the few domestic machines available. Like many other computers at that time, it had to be housed in an air conditioned room to ensure that it would function properly.

The growth in acceptance of the SP Alarm System was paralleled by an increase in the number of bank robberies. Throughout Japan, from 1960 through 1964 there were only eight bank robberies or 1.6 per year. From 1965 through 1969 the annual average increased to 11.4 with the absolute number for the period being fifty-seven. This trend

6 Control Centre, Tokyo, 1967.

would continue. From 1970 through 1974 there were 188 robberies or 37.6 annually. During the next five years the annual average rose to 56.8 and bank robberies for the period totalled 284. In 1980 there were as many as 156 bank robberies.[5]

The need for foolproof security was evident from an early date. The introduction of the SP Alarm System set the stage for Nihon Keibi Hosho's transformation from a labour reliant company to a security focused communications network business. As the range of users of the system expanded to include factories, museums, art galleries, and nuclear plants, the SP Alarm System changed forever the face of security in Japan.

Improvement in the external security infrastructure was a beginning. For Iida what was now important was to create an accompanying lean, effective corporate constitution. At every opportunity he discussed with his employees the fact that the concept of the SP Alarm System should not only be considered as a 'mechanical' improvement but rather as an important step towards the creation of an integrated hardware/software holistic system. Through the synthesis of technology and human skills, Nihon Keibi Hosho would be able to gain the economies of scale and scope required to maintain the corporation's role as a pace setter in the security industry. Through entry into the field of machine security, the company would be able to realize the low cost and product/service differentiation that would create a comparative advantage within the industry and eventually lead to 'enterprise differentiation'.[6]

ACCUSATION OF ESPIONAGE

At the same time that Iida was pursuing the development of the SP Alarm System and he and his company were becoming well-known through the television series 'The Guardman', a magazine article hinted at a link between Iida and international communist espionage activity. Once in print, the rumour spread quickly.

The accusation both dumbfounded and amused Iida but it took its toll on the business. Existing clients questioned his integrity and potential customers hesitated to sign contracts. The term 'spy' elicits a fierce negative reaction among Japanese. Once branded, a tarnished image is hard to shake.

Japan's geographical proximity to the eastern fringe of communist block countries, the presence of American military bases in Japan, and the nation's political and economic ties to capitalist countries combined to foster a variety of espionage activities including the use of Japan as a base for communist spy activities against third party nations. The purpose of the espionage was mainly to gather information with regard to Japan's political, economic, diplomatic, defence, and high technology activity; the military activity of the United States; and the political and military activity of South Korea. Japanese law did not allow for the direct control of spy activities. Spies could be arrested only when they broke laws within the existing criminal code, a situation that made the conduct of espionage activities in Japan relatively easy.[7]

Between the mid-fifties and the mid-sixties a number of *Kyokusa Boryoku Shudan* or 'extreme left violent groups' emerged in Japan. These groups were intent on triggering a socialist/communist revolution which would destroy democracy through terrorism and guerrilla activities. Since their first appearance in 1957, these organizations had formed into a variety of splinter groups engaged in the destruction of people, residential neighbourhoods, companies, and transportation facilities.

By the time the SP Alarm System was becoming known in Japan, membership in these terrorist groups was peaking at 53,500.[8] As their activities intensified they also became more violent with widespread use of weapons ranging from sticks, stones, bamboo spears, and iron pipes to guns, explosives, poisons, and gasoline bombs. It was within this social setting that the rumour of Iida's espionage activity found fertile ground.

The origin of and explanation for the accusation have never been clarified. Iida suspects that dislike of the initial heavy foreign investment in the company may have been a contributing factor. Or it may simply have been jealousy on the part of a business rival. The president of a newly emerged security firm and the publisher of the magazine which first carried the story were friends.

Another explanation Iida offers is retaliation on the part of reporters he tended to ignore when they wanted to discuss his company's success. Finally he yielded to an interview with the prestigious *Asahi Shimbun*, one of Japan's leading liberal daily newspapers. Jealousy may have set the rumour mill in motion. In due time, lacking any basis, the rumour evaporated.

INTERNAL THEFT

By 1966, Nihon Keibi Hosho was in the initial phase of becoming a nationwide security operation, the installation of the first SP Alarm System at the Mitsubishi Bank had been completed, and the dust had settled after the espionage accusation. Iida was beginning to breathe a sigh of relief when another crisis struck. A security guard at Isetan department store was convicted of stealing jewellery valued at a total of seven million yen.[9] Then a guard broke into a supermarket safe and pocketed the contents. Four more internal robberies occurred within the next month.

With Nihon Keibi Hosho already in the public eye through the television drama it is not surprising that the robberies were given extensive media coverage. Nihon Keibi Hosho even became the butt of jokes by stand-up comedians. At this time Iida's son was attending an expensive Tokyo private school. Hearing of the robberies the teacher feared that the company would not be strong enough to survive the series of setbacks and inquired of Iida whether or not he wished to continue sending his son to the school. Saddened that even his small boy had to share in the company's disgrace, Iida strengthened his resolve to make amends.

As Iida expected, potential clients were reluctant to sign contracts. This problem was especially troublesome in those newly established branch offices located in areas where Nihon Keibi Hosho's reputation had not had a chance to penetrate. Many clients about to do business with the company withdrew. Of even greater concern to Iida was the loss of trust in the company on the part of existing customers. He immediately contacted and/or visited all his clients to apologize and to explain personally what steps he would take to prevent the recurrence of such incidents.

His confidence shaken by the robberies, Iida visited his father. After staring at his son for sometime without speaking, the older man drew four characters on *tanzaku*, a rectangular strip of Japanese calligraphy paper. They read: *Shoshin Wasuru Bekarazu.* This phrase is part of a longer quotation from Zeami, the founder of Noh drama. Its meaning is 'Do not lose sight of your initial determination'. The impact of his father's words was sufficient to restore Iida to his initial determination to contribute to society through the provision of a high quality security service to customers, in a way beneficial to employees and shareholders. More than twenty years later the *tanzaku* still hangs in Iida's office.

There is a strong identification among Japanese employees with the companies that employ them. Iida was concerned that the shame brought on Nihon Keibi Hosho by the robberies would affect the lives of his employees and their families even to the extent of many tendering their resignation. Using the pretence of site inspection, each night for several weeks Iida visited a number of guards in an attempt to boost morale.

Anticipating comments from the guards like 'We are ashamed', 'How can we cope with the embarrassment of the situation?' he was prepared to implore them to be patient, but the reality differed from what he had anticipated. Everyone to whom he spoke pledged their support to Iida and the company and vowed to work together to rebuild Nihon Keibi Hosho. No one resigned. He was heartened by their response. 'In retrospect', says Iida, 'the robbery crisis fostered a spirit of unity within the company.'

Iida contemplated what course of action would best steer the company through the crisis. Some employees suggested that instead of climbing upward without pause the company's progress should be similar to the process of climbing stairs – one flight at a time pausing on each landing. The company was now at a landing and should take time to deal with the problem of employee theft and not proceed upwards until the necessary adjustments had been made to prevent recurrence.

Iida did not share this opinion. If the company had been on a firmer footing with the necessary reserves to see it through setbacks then such an approach might have been appropriate. But Nihon Keibi Hosho was still very young. Taking time out to recuperate could bankrupt the company, especially now that there was firmly rooted competition in the security industry. Iida made up his mind to repair the damage rendered by the thefts as the company ran at full speed. He urged his sales department to work as hard as possible to obtain contracts. From this experience he became convinced that in times of crises, if top management shows the direction and leads the way, employees will follow.

At the same time, Iida felt, major changes in personnel management were desperately needed. By 1966, the time of the robberies, Nihon Keibi Hosho had more than 600 employees. Although he had personally interviewed each of them during the recruitment process, Iida felt that the company's screening and employee education procedures were deficient. He decided that it would be in these areas that he would focus his attention to prevent a recurrence of employee crime.

Iida felt personally responsible for having hired men who would indulge in criminal activity. He feared their behaviour was a manifestation of a lack of professional ethics throughout his work force. In order to counteract this he embarked on a series of informal meetings with his security guards throughout the country, asking them to express their true feelings, dissatisfactions, and aspirations directly to him. These intense discussions at the grassroots level afforded Iida the opportunity to share his managerial ethics, corporate mission, and his views on societal responsibility. Called *Shasho o Mamoru Kai* or 'Living up to the Corporate Logo', these corporate-wide 'man-to-man' talks were institutionalized and became the basis for later employee training and education seminars.

At the Isetan Board of Directors' meeting the question of whether or not to continue with Nihon Keibi Hosho's security service was discussed. Opinion varied but in the end the decision was made to retain the service based on the knowledge that the company's intentions had been honourable from the outset. Board members felt that being young and somewhat inexperienced the company deserved a second chance. They knew that cancellation of Isetan's contract could destroy Iida's hopes and dreams. Other companies using Nihon Keibi Hosho's services watched and waited for Isetan's decision. When Isetan decided in favour of Nihon Keibi Hosho these companies followed suit.

The robbery crisis was thus resolved without causing permanent damage to the company. It was an incident from which all employees could learn and Iida continues to discuss it with his staff at every opportunity. He stresses the fact that a crisis may heighten feelings of employee solidarity but that these feelings need to be present in the workplace at other times as well. Each corporation has a mission, Iida believes. In order to realize its accomplishment corporate members must maintain the tension that is automatically present in a crisis situation, constantly re-examining the tasks which must be completed. Even unfortunate events became resources upon which Iida could draw to further his entrepreneurial goals.

The accusation of espionage, the internal thefts and the daily running of the business together created a spiritual crisis for Iida.

> It was as if my heart had become a dried out rubber ball. Constantly being squeezed, it had lost its resilience. I knew I had to alter the situation. For three months I trained my mind to think of myself as resilient, always able to bounce back. Every part of my body felt as if it were restrained. I deliberately tried to relax, to release the tension. Gradually I began to feel that through mental training I could create a positive, rather than negative, cycle. Now this process happens almost automatically and I can maintain a positive outlook.

With this positive outlook Iida challenged the development of a more dynamic corporate culture. His resilience was infused to the members of his work force and the company continued to grow. By the end of 1968 Nihon Keibi Hosho had established twenty-four regional branches and sales offices in major centres throughout Japan. Through this nationwide effort the company was changing workplace attitudes towards security and safety with on-site guards playing an important role in heightening awareness of preventative measures.

Saburo Saito was assigned to guard duty at a Tokyo chemical plant when the company, dissatisfied with its original security guard service, signed a contract with Nihon Keibi Hosho in 1968. Says Saito:

> I was concerned about my ability to handle properly this assignment. I thought perhaps it should be given to someone with some expertise in dealing with chemical fires as I perceived the risks in this area to be high. Initially, corporate attitudes towards the security service were poor. The previous security company had not done a good job and although management had changed companies they didn't expect much improvement. My first task was to obtain their trust. To do this I followed Nihon Keibi Hosho's four operational cornerstones: sincerity, responsibility, alertness, and service.

Initially, the attitude of all but a few employees of the chemical plant towards safety was poor. Cigarette butts were carelessly discarded, electricity left on when it should have been shut off, and windows and doors left unlocked when employees departed for the day. Despite this sloppy behaviour the client's corporate daily report always indicated that everything was fine.

I accumulated statistics and reported the situation to the chemical company. The employees' attitude gradually changed and their behaviour became more responsible. Eventually common sense safety practices became habitual.

In addition to security tasks Nihon Keibi Hosho guards – and Saito is one example – were sometimes asked to assist with the maintenance and operation of factory machines. The economy was booming and labour shortages were critical in the late 1960s.

I sometimes wondered how the chemical plant could have managed without me considering all the non-security related tasks I performed. Many other Nihon Keibi Hosho guards felt the same way. We perceived ourselves as vital parts in the clients' large corporate machines – machines that would grind to a halt were we to withdraw our services. I may have had an exaggerated view of my own importance but, in any case, it stemmed from the in-house education I received with Nihon Keibi Hosho.

THE ALARM SYSTEM PROVES ITSELF

While the human-based security system continued to prove effective, Nihon Keibi Hosho was expanding rapidly in the area of machine integrated security. Despite several setbacks, within three years of the company introducing a machine integrated security, the system had conclusively proved its worth. The robberies convinced Iida that the time had come to curtail the company's reliance on guards and expand the use of the machine integrated security system. The event that highlighted its effectiveness was the capture of a wanted criminal.

When the alarm in the Hitotsubashi Business School in the Yoyogi district of Tokyo was tripped early on the morning of 7 April 1969, it registered in Nihon Keibi Hosho's control centre and a company patrol car was immediately dispatched to the site. The police were notified as well. The first on the scene was the guard, Toshimi Nakaya. External examination of the premises revealed that someone was still inside.

Nakaya entered the building and shouted: 'Who's there?' The

surprised burglar, hidden behind the reception counter, drew a gun on the guard. With Japan's tight gun control laws, armed robberies are rare and Nakaya assumed the pistol to be a toy. He approached the intruder and knocked the gun from his hand with his nightstick. A fight ensued and the robber managed to retrieve his gun. He fled the premises, firing at Nakaya as he did so. Fortunately he missed.

A second security guard, Masao Sasaki, arrived in time to see the suspect flee and gave chase.

> I had been a track athlete in high school and had confidence in my ability to overtake the intruder. But he was desperate and ran even faster than I did. In the darkness I lost him as he headed into a deserted area with many parks, the large Meiji Shrine, and a sizeable athletic field.

The area was cordoned off by police who had received a detailed description of the man from the guard. Within a short time they had apprehended him as he tried to blend in with the early morning strollers. He was identified as 19-year-old Norio Nagayama, wanted on a number of murder charges. In the early hours of 11 October 1968 Nagayama had shot and killed a Sogo Keibi Hosho security guard at the Tokyo Prince Hotel. Three days later he shot to death a night watchman at the Yasaka Shrine in Kyoto. This was followed on 26 October by the daytime shooting death of a Hakodate taxi driver in Hokkaido and on 5 November he shot and killed a taxi driver in Nagoya.

All of the shootings were apparently without a motive. Although the third and fourth had occurred in conjunction with robberies, the amount taken was so small as to be nearly insignificant. The murder weapon, still on Nagayama's person when he was taken into custody, was a .22 calibre revolver stolen from the American naval base at Yokosuka. Nagayama had shot each of his victims through the head at close range. He had eluded an intensive police search until he was finally apprehended with the help of Nihon Keibi Hosho's alarm system. In the early years of his imprisonment Nagayama published two books explaining the circumstances which led to his criminal activities.[10]

During much of his trial Nagayama remained silent. However, in June 1970 he offered an explanation for his behaviour based on the considerable reading he had done while in jail. His criminal acts, he believed, were rooted in poverty and ignorance. Speaking in English and quoting a passage from William A. Bonger's *Criminality and Economic Conditions*, Nagayama stated: 'poverty . . . kills the social sentiments in man, destroys in fact all relations between men. He who is abandoned by all can no longer have any feelings for those who have left him to his fate.'[11] In 1979 Nagayama was sentenced to death. He appealed and in 1981 the sentence was reduced to life imprisonment. In 1983 the crown prosecutor appealed the 1981 decision and in 1987 it was reversed. The Supreme Court sentenced Nagayama to death. Nagayama appealed but his appeal was unsuccessful. He presently awaits execution on death row.

Nakaya and Sasaki received Metropolitan Police Commissioner's Awards for their part in the apprehension of the murderer. Iida promoted both guards and presented them with monetary rewards in recognition of their bravery. Despite this acknowledgment, Nakaya's behaviour on his arrival at the business school came under scrutiny. This led to a re-examination of the company's training system. Subsequently, greater emphasis was placed on appropriate crisis response. Training routines underlined the importance of guards taking every precaution to protect their own lives, remembering that responsibility for the apprehension of an intruder lay with the police.

At the awards ceremony Iida congratulated all those involved in the Hitotsubashi Business School case – Nakaya, Sasaki, their immediate supervisors and those in the control centre at the time of the incident. Later, he took every opportunity to remind his staff that the company should not rest on its laurels after this success. It was necessary to strive actively towards the prevention of crime in order to protect society. While Nihon Keibi Hosho's involvement in the apprehension of a wanted murderer had been dramatic, Iida pointed out that company guards were constantly identifying factors which, if left unchanged, could result in accidents or crime.

Company records show that in 1966 alone 4,543,332 potentially dangerous situations were identified at guarded sites. Included in these were such things as 1.2 million doors, windows, and gates which could not be locked because the keys were either damaged or lost; 20,000 open safes containing valuables; and 114,000 defective fire extinguishers. As well, guards identified 47,000 unauthorized strangers and escorted them from the premises, extinguished 495 fires, and assisted in life saving emergency activities 377 times.

The fact that Nagayama's first victim had been a security guard and that security guards played a key role in his capture served to increase public awareness of the social function of professional guards. If an alarm system had not been in place at the Hitotsubashi School of Business, Nagayama might have remained at large for a considerably longer period and killed more victims. This incident not only improved the public image of the security firm, it also increased recognition of the effectiveness of machine integrated security systems as a means to counteract crime. Many instances of the SP Alarm System contributing to the arrest of wanted criminals followed.

Nihon Keibi Hosho's role in apprehending an accused murderer was widely reported. One of the difficulties confronting the firm in the early period of societal adoption of the SP Alarm System was the ability to make the usefulness of the product known. The adoption rate of an innovation depends on its communicability – the extent to which the impact of an innovation can be observed by others.[12] The usefulness of security service is measurable in terms of the absence of crime but this absence was difficult for consumers to gauge. The Nagayama incident was a decisive event which made the usefulness of the SP Alarm System visible in the public eye.

Nihon Keibi Hosho's role in the apprehension of a dangerous criminal was also acknowledged internationally. At the thirteenth biannual meeting of Ligue Internationale des Sociétés de Surveillance held in Munich in September 1969, the company was awarded the Ligue's gold medal for distinguished public service. Two other corporations, Pinkerton's Incorporated of the USA and Total Security Limited of England, were similarly honoured that year.

Another widely reported event which contributed to the increase in public awareness of the need for professional security services occurred on the morning of 10 December 1968. A truck transporting cash belonging to the Kokubunji branch of the Nihon Shintaku Bank was stopped by a young man posing as a motorcycle policeman as it travelled through Fuchu city, on the outskirts of Tokyo. Flares were set up and the 'policeman' instructed the four bank employees to get some distance away from the truck as a report had been received that it was carrying a bomb. They did so and the 'policeman' hopped in and drove away with three hundred million yen. He was never apprehended.

By early 1969 the number of Nihon Keibi Hosho employees had increased to three thousand. Iida felt the time had come to commission a company song.[13] Japanese businesses often set their corporate values to music, creating a symbolic representation of the company that can be reaffirmed by employees through song. Hachiro Sato wrote the words and Hideaki Yasu composed the music. At the same time, an *aishoka* or 'favourite company song' to be sung on less formal occasions was commissioned. The songs were ready for presentation at the August branch managers' meeting. Subsequently, tapes of the songs sung by the popular singer Koichi Miura accompanied by the

Nippon Victor Orchestra were sent to each branch office so that all employees could learn them. In many of the branch offices the song would be sung daily following the morning briefing as well as on ceremonial occasions and at times of company recreational travel.

6 Internal and societal adjustments

MANAGERIAL EXPANSION

Nihon Keibi Hosho's growth necessitated expansion of its management component. In order to implement his vision for the company's development Iida needed personnel capable of accomplishing his strategic objectives. There was a limit to what Iida and Toda could accomplish on their own. While the company had many committed individuals willing to devote themselves wholeheartedly to immediate planning and accomplishment, it lacked those who could participate in formulating and achieving general business directions within Iida's grand design for the corporate mission. Iida needed to develop committed executives who could work with him towards the implementation of his vision.

The primary source of managerial candidates was university graduates. The company's small size and short history, however, made recruitment difficult as the majority of new university graduates gravitated towards larger, established companies. It was not until 1970 that the company was able to attract its first graduate directly from university and even then it was only one. However, even university graduates require years of nurturing before they are ready to assume the type of role Iida envisaged for his managerial staff. In an effort to fill the void at the managerial level created by the company's rapid growth, Iida looked first to friends from high school and university days who were in established positions with other companies. He was assisted in this endeavour by the media coverage of Nihon Keibi Hosho's part in Norio Nagayama's arrest which led to a number of congratulatory calls from friends.

Kei Kimura had attended junior high and high school with Iida. After graduating from Meiji University in Tokyo, Kimura had found employment with Nippon Denshin Denwa Kosha, the government-owned telegraph and telephone corporation which was later privatized to become NTT (Nihon Telegraph and Telephone Corporation). He was involved in the guidance and inspection of routine activities in the approximately 100 telegraph and telephone offices throughout Tokyo. When Kimura read in a weekly magazine about Iida and the arrest of the wanted gunman he telephoned to offer his warm congratulations on the company's success. Iida explained to Kimura that with the SP Alarm System's dependency upon Nippon Denshin Denwa Kosha's exclusive circuit he was looking for someone to employ who could facilitate the link between the two companies and who had a good knowledge of communication networks. Ideally, this person would have been employed previously by Nippon Denshin Denwa Kosha.

Kimura agreed to help Iida in his search but six months passed without success. There was no one willing to relinquish the security of employment with a large, well-established government corporation and chance a move to a smaller, lesser known company. This was not surprising as changing employers was not a common practice in Japan, although its popularity has increased in recent years. While personnel flow from business to government service is practically unknown, retiring senior government officials commonly accept managerial positions in private corporations. This practice is known as *amakudari*, literally 'descending from heaven'. Kimura was a long way from retirement but when Iida suggested that he might consider the move himself he accepted the challenge and began his career with Nihon Keibi Hosho at the end of 1969. In 1985 Iida appointed him president of the associated company Secomnet and two years later he assumed the presidency of SECOM, positions which he held until his death from heart failure in June 1989 at the age of 55.

Still searching for management personnel Iida approached Kenichi Komine, a friend from high school and university days who had graduated from the Faculty of Political Science and Economics at Gakushuin University the year after Iida. He was working for the Nippon Kangyo Bank which, in 1971, would merge with the Dai-Ichi Bank to become Japan's largest bank. Iida described the prospective job as interesting and exciting, characteristics verified by Kimura when Komine sought out his opinion on his own move. The bank offered security but Komine had become bored with the routine nature of the work. He wanted a new challenge to test his ability. In August 1970 he joined Nihon Keibi Hosho.

The corporate atmosphere surprised him at first. Compared to the bank, the atmosphere was lively and vivid. Everything moved quickly with very little red tape. Iida made decisions swiftly and displayed complete trust in his employees, leaving each one free to work out the best way to do the job at hand. Says Komine: 'I put everything I had into my job. It was exhilarating.' In 1990, eight months after Kimura's death, Komine was appointed president of SECOM.

Hisaki Shimosato, Iida's friend from university days, was the third person to be recruited at this time. Upon graduation he had found employment with the Mitsubishi Automobile Corporation. In 1969 he was the company's top salesperson. Over the years he had had some dealings with Nihon Keibi Hosho, supplying the company with a number of its patrol cars. He, too, read about the capture of the gunman and phoned Iida to congratulate him. He was curious to know how the alarm system worked. In February 1970, knowing that Shimosato was interested in the company's work and that he had respect for his *sempai*, Iida asked Shimosato to join the company. Shimosato was tempted but was unable to comply with Iida's request that he begin immediately. He was committed to the completion of an advertising and promotion campaign for a new car that Mitsubishi was bringing on the market in April. In June when his obligations to Mitsubishi were fulfilled he submitted his letter of resignation and joined Nihon Keibi Hosho.

> When I joined Nihon Keibi Hosho the SP Alarm System was enjoying a rapid increase in popularity. Iida gave me a detailed explanation of the concept on which it was based. As I listened I was struck by the similarities between the company's security architecture and the dynamic network strategy of American football where team members are linked through coordinated and integrated systematic team actions.

Thus, Kimura, Komine, and Shimosato, all of whom went on to become senior executives with the corporation, joined the company not because of its reputation and standing, but rather because of Iida's magnetism. Among Japanese, importance is placed on the relative age of individuals. Being slightly younger than Iida and having attended school with him, the men had an inherent respect for their new boss which deepened as they came to know him better in this new capacity.

It has been pointed out that corporate entrepreneurs, while not necessarily outdistancing other people in their creative abilities, are capable of sharing their visions with their colleagues and building coalitions among them. The result is a working team whose members back the entrepreneur's ideas.[1] The camaraderie that developed among Iida, Toda, and Osada in the early stages of the company's growth contributed much to the creation of a corporate culture in which Iida was securely at the helm.

Iida also recruited individuals whom he had come to know through his work. Following the capture of the assassin, Norio Nagayama, the SP Alarm System received wide publicity. Iida wanted to capitalize on the growing public awareness of the efficacy of security systems and use that opportunity to penetrate the market frontier. As well, the growing business required an influx of qualified, vibrant personnel. In order to attract these people it was necessary to launch an appealing, wide-reaching advertising campaign. Additionally, intra-corporate communication needed to be improved. Previously, external and internal public relations had been conducted in an ad hoc manner. Iida felt that the time had come to hire a public relations officer to coordinate and expand company efforts in this area.

Zenjiro Kato was selected for the job. He had worked for three and a half years for Ad Dentsu, a subsidiary of Dentsu, now the world's largest advertising corporation. During that time he had handled Nihon Keibi Hosho's advertising contract, among others, and had been involved in nationwide recruitment advertising campaigns for other companies. At the time the decision was made by Dentsu to transfer Nihon Keibi Hosho's advertising activities from Ad Dentsu to the parent corporation, Kato resigned and moved to *Iwate Nippo* where he worked for eight years in the newspaper's strategic planning and public relations departments. When he was invited to join Nihon Keibi Hosho in 1970 he had been in charge of the *Nihon Keizai Shimbun* (newspaper) advertising agency for a year. Kato recognized the direct correlation between the extent to which the inherent utility of a corporation's products and services was understood by the general public and the degree to which corporate development could occur. In 1972, two years after Kato joined Nihon Keibi Hosho, the company established a public relations division under his directorship.

The incident at Hitotsubashi School of Business served not only to increase sales of the SP Alarm System, but also helped secure the human resources necessary for the company to progress. The method of management personnel selection introduced when Shinichiro Osada joined the company in 1966 was continued with the next three recruitments, and thereafter, to the benefit of the company. By 1990 excluding Iida and Toda, only one of SECOM's twelve top executives had not been recruited from other corporations. Iida comments:

> I am glad many of our executives and other employees gained experience elsewhere. A company filled with people who have known only one corporate environment, who have grown up together, speak the same language, and think in the same way

is, in my view, narrowing. People from different backgrounds meet and mesh like different gears, propelling the corporation forward.

While Iida continued to handpick his managerial staff, the process did not exclude Nihon Keibi Hosho employees. One of those who climbed the ranks from within and came to play an important role in the company was Shohei Kimura. Kimura had hoped, upon graduation with a degree in French literature from Doshisha University in Kyoto, to pursue his interest in play and film production in France. To do this he required money and, so, in 1967 he found employment with the Sapporo branch of Nihon Keibi Hosho. Although he was among only five of 130 applicants hired, he had no intention of making this work his life-long career. In the summer he was contemplating quitting the company and pursuing his original plan when Iida visited the branch and the two men met for the first time.

Kimura was immediately attracted to Iida's vibrant personality. Shortly after this, along with the other university graduated employees, he participated in the company's first managerial training programme, held in Tokyo. Here, he had the opportunity to hear Iida speak about his values and philosophy and his future plans for the company. Subsequently, there was a dramatic increase in Kimura's interest in working for Nihon Keibi Hosho and a gradual waning of his dream of studying abroad.

Iida's ability to motivate people is typical of great leaders in all fields worldwide. Despite their many differences people are, at their core, remarkably similar in their hopes and fears, their need to feel loved. Iida has an acute sense of what people are about. In the words of Michael Kaye, president of SECOMERICA, SECOM's American holding company, 'Iida makes people feel very filled up, very understood, very loved and beloved, and that is incredibly motivating to people.'

Recognizing Kimura's managerial potential, in the autumn Iida presented him with a challenging opportunity. The company's branch in Tsu city in Mie prefecture was in the throes of a severe financial crisis and required rebuilding under the direction of a new manager. Iida offered Kimura the job. Kimura accepted the challenge. Relatively new to the company and with a background in literature, Kimura had little knowledge of finances and balance sheets. However, he studied the situation carefully, learned all he could about financial matters and within a year and a half had transformed the failing Tsu city branch into a profitable operation.

With the branch back on its feet, Kimura turned his attention to the independent study of management in general and business strategy in particular, and formulated in his own mind the future direction of Nihon Keibi Hosho. In 1970 he wrote a report entitled 'Nihon Keibi Hosho – A Five-Year Plan' and submitted it to Iida. Herein, he emphasized that the telecommunication circuit used for the SP Alarm System was a primary resource upon which future corporate expansion should be based. Several years later Shohei Kimura would become an executive director of the company.

Iida was already contemplating the development of this resource to its full potential, through corporate expansion into a 'larger network service business'. Knowing that future development should be in the direction of computer controlled machine-based security, and to create the necessary cost-effective communications network, Iida established a new SP Alarm control centre in the Harumi district of Tokyo. This became Nihon Keibi Hosho's communications hub. Eighteen other centres were established throughout Japan to service those regions where Nihon Keibi Hosho had established branches. In major cities these centres were supported by strategically located depots each with approximately ten patrol cars for the rapid dispatch of guards to a problem site. Metropolitan Tokyo, for example, was serviced by nine depots each responsible for a 4-kilometre radius.

The depots were staffed with experienced guards. Toshimi Nakaya and Masao Sasaki, the two guards associated with the capture of the wanted criminal Nagayama, were assigned to supervisory positions at the Senju and the Ebisu depots respectively. At about the same time, Iida founded Nikkei Densetsu Corporation, a subsidiary of Nihon Keibi Hosho, to handle the installation of security systems. Today the company is known as SECOM Service. This subsidiary and the subsidiary SECOM Maintenance together form a support structure for SECOM's services.

ON-SITE SECURITY

In the late sixties, Nihon Keibi Hosho offered three types of security service: permanent, on-site guards; periodic premises check guards; and the SP Alarm System. At this time the first two types of service comprised 95 per cent of total sales. The SP Alarm System accounted for only 5 per cent.

In addition to businesses, one of the early primary users of on-site guards were temples. In 1968 there was a rash of thefts of temple treasures in Kyoto, Japan's cultural capital and home of many of its oldest and most historic religious sites. Numerous temple buildings and the treasures within such as statues, paintings, scriptures, and ceremonial equipment were nationally designated as important cultural properties. Many of these items were accessible to the public and closely linked to ritual activities. In addition, the temple gardens were fragile cultural expressions requiring protection from the masses of people entering the temple precincts daily. For priests and temple administrators alike, finding effective security was essential.

Nihon Keibi Hosho considered the protection of important cultural properties in temples in Nara and Kyoto, the first and second capitals of Japan, to be an important social responsibility. In order to fulfil this responsibility a special temple patrol was organized and trained. Towards the end of 1968 Saiho-ji, commonly known as 1 oke-dera or Moss Temple because of its beautiful moss-covered grounds, was one of the earliest temples to turn to Nihon Keibi Hosho for security guard service. In addition to the problems shared by all temples, the priests at Saiho-ji were concerned about the preservation of the many and varied mosses which carpeted the ground. Visitors had a tendency to remove samples, presumably for transplantation to their own gardens. Pleased with the results when Nihon Keibi Hosho guards were hired, the priests recommended that the Association for the Preservation of Ancient Culture in Kyoto assess the quality of guard service provided by the company. When Iida was invited to meet with members of the Association in Kyoto he took the opportunity to explain the role of the private security corporation in the protection of cultural heritage. The Association gave Nihon Keibi Hosho a favourable evaluation and, subsequently, many Kyoto temples concluded contracts with the company.

Soon Nara temples were doing likewise. Among them were Kofuku-ji, Yakushi-ji, Todai-ji, and Toshodai-ji, all temples dating back to the seventh or eighth centuries. Kofuku-ji presented a particular challenge. Under *Haibutsu-kishaku*, the policy of the Meiji government to eradicate foreign-introduced Buddhism and restore native Shintoism in order to heighten nationalism, many temple precinct walls had been destroyed and the grounds incorporated into public parks. This was the case with Kofuku-ji. A century later its precincts had become a meeting place for young lovers, short cuts for urban commuters, and a place to smoke, visit, and relax for neighbourhood residents. It was extremely difficult to regulate entry to the temple. The problem was compounded by the fact that the temple housed a museum of national treasures that was open to the public year round.

Shinto shrines, too, provided a security challenge. Because of

Shinto's close connection with the Imperial system, some important shrines, including Ise, the holiest of all Shinto shrines, became targets of terrorist attack in the late sixties and early seventies when feelings against the imperial system were running high. The provision of professional security lessened considerably the risk of damage and theft.

One of Nihon Keibi Hosho's major contracts in 1969 was with Kikkoman, the nation's leading soya sauce manufacturer. The company provided over-night security guard service for Kikkoman's several factories in Noda city in Chiba prefecture. Throughout the night guards would circulate among the factories, a fact that pleased residents in the area as they felt that the guards travelling from one factory to the next helped to ensure the safety of the entire area. Another service Nihon Keibi Hosho provided at this time was the provision of a nine-member female security patrol at Kansai International Airport in Osaka.

The Tokyu Group, headquartered in the Shibuya district of Tokyo, was another large client which signed on with Nihon Keibi Hosho in the late sixties. Its largest component is the Tokyu department store which encompasses within its confines in the heart of Tokyo the convergence of four railway lines. One and a half million commuters pass through this area daily. On-site security was required round the clock.

While most of Nihon Keibi Hosho's clients were in urban centres, some were in the countryside. Fujita Mujinto Paradise was one such client. The company had turned a portion of the island of Naoshima in the Inland Sea into a resort. Guards were hired seasonally to secure the campsites, beaches, and botanical gardens. Market expansion in on-site security was the outcome of the company's conscious effort to expand the market frontiers.

In addition to permanent on-site security there was an ongoing need for this type of service on a temporary basis at special events. Nihon Keibi Hosho had been involved with this type of security on a large scale at the Olympic Games and at the Harumi exhibition. In the late sixties the company signed a contract to provide security at Expo '70, held in Suita city on the outskirts of Osaka from mid-May to mid-October. Participated in by seventy-seven countries, the exhibition attracted a total of sixty-four million people.[2] From the beginning of the construction of the pavillions in August 1968, Nihon Keibi Hosho was one of the providers of more than a thousand on-site security guards.

In addition to seasonal contracts Nihon Keibi Hosho had other temporary on-site contracts, some of which were with construction companies. It was at a construction site that the second death of an on-site security guard occurred. In 1969 the Merchandised Mart Building was under construction in Osaka. Kenji Nakagawa was making his rounds of the site on the morning of 27 June when a steel girder fell on him killing him instantly. Nihon Keibi Hosho and the construction company, Takenaka Corporation, together provided for the funeral expenses and arranged for monetary compensation to Nakagawa's parents. He was not married.

By the late sixties recruitment of personnel to fill the on-site security guard requirements of Nihon Keibi Hosho's ever growing number of contracts was becoming increasingly difficult. In the booming economy of the time blue collar workers were in high demand while at the same time the pool from which they could be drawn was shrinking as more and more Japanese continued on to secondary and post-secondary education. Advertising for guards in the urban print media was not producing sufficiently rapid results so Nihon Keibi Hosho launched an active recruitment campaign throughout the Japanese countryside. Staff from the personnel department were dispatched to rural communities where Nihon Keibi Hosho's advertisements were run in the local newspaper. Temporary recruitment stations were set up at the local newspaper offices where respondents could be interviewed and hired. The success rate varied but generally this approach helped alleviate the shortage of human resources.

ABOLITION OF PERIODIC SECURITY CHECKS

While the services provided by on-site guards at temples and special event sites could not be replaced by electronic alarm systems, Iida knew that expansion of the machine security system with a corresponding reduction in periodic premises checks would lead to improved service at reduced cost. Rising labour costs and training expenses were forcing the price of security skyward. At the same time the efficiency of the service was restricted by several factors: the area to which any one security guard could give adequate coverage was limited; imperfections in human attentiveness resulted in errors; despite intensive training, guard response to particular situations was variable. Iida met with his executives to review the situation and ascertain their thoughts on the matter in the summer of 1970.

A heated discussion of the merits of the periodic security check versus the electronic security system ensued. The hard work that had gone into marketing the periodic security check service was manifest in the company's 4,000 active contracts for this service. The consensus was that the service should be maintained and that the SP Alarm System should be a supplementary service only.

The traditional human-based security service, Iida felt, could be easily copied by other companies. More importantly though, the quality and scope of the security it offered could be substantially improved through the addition of electronic surveillance equipment, freeing personnel for other necessary and challenging activities. Iida's business strategy has always been based on the assumption that current technology and services are continuously in the process of becoming obsolete. To this end he is constantly asking the question 'What should our business be?' and planning the systematic shedding of the out-worn. This frees finite resources for redirection towards new ventures.

Iida is well aware of the importance of growth but is determined that it will occur not for its own sake but only as the result of the company 'doing the right things' and becoming better in the performance of the services it offers. Iida lives by the principle that no activity will or should stay the same forever – that any practice will be abandoned as soon as it can be improved upon. He has always known that for true growth to occur it would occasionally be necessary to give up activities which no longer offer customers the best possible service nor contain within them realistic potential for growth. The periodic security check was one such activity.

Opposition to the replacement of the periodic security check with machine security was grounded not so much on the possible loss of profitability, but rather on the belief that the change would clash with the organization's traditional mission and purpose. The clash of new ideas with the old is a common problem in corporate development. 'Old ideas represent the "ghost in the machine", ready to haunt those who strive to introduce something new. They trap us in ways we don't even understand in the same ways that old routines imprison us.' [3]

Iida had made up his mind. He understood that corporations live by change. The next steps were evident to him. He announced that the periodic security check service would be 'abolished' and replaced by the machine security system. His executives were stunned. No one had contemplated such a drastic measure. Iida's decision to abolish the periodic security check meant adjustment in sales activities. Customers of the service had to be notified of the impending change and their contracts adjusted accordingly. This decision resulted in a substantial increase in contracts for the SP Alarm System. By the end of 1971, contracts for the system had tripled those of the previous year, numbering 5,116. The company's sales totalled 6.3 billion yen. The remarkable success of the system in the marketplace helped to allay residual employee opposition to the shift away from periodic security checks. Employees no longer needed to carry out on-site duties were retrained and assigned to other areas within the corporation. In addition to the single private circuit SP Alarm System, in October 1971 Nihon Keibi Hosho introduced the SP Alarm Pack using public telephone lines.

Iida was not one to wait for the opportune moment to launch a new venture; he preferred to start immediately once the decision had been made and work out problems as they arose. With the increased sales of SP Alarm Systems it was necessary to establish more and more depots so that every client would be in close proximity to a depot from where beat engineers could be dispatched. Beat engineers were newly employed individuals as well as security guards freed from periodic security check duty and retrained for the new position.

With contracts increasing at the rate of four to five hundred a month, it was hard for depot expansion to keep pace. To bridge the gap, Iida took an innovative approach. In 1972 he introduced twenty-three mobile depots into areas where contract sites were widely scattered or where depot establishment was behind schedule. Mobile depots were used in Tokyo, Yokohama, Shizuoka, Nagoya, Osaka, Kobe, and on Shikoku island. They were also used at the 1972 Sapporo Winter Olympics.

Designed to serve as mobile depots, each Mitsubishi built mini-bus contained short wave communication facilities, office space, and a depository for the keys of each of the secured premises in the depot's territory. When an emergency signal was received by the control centre the appropriate mobile depot was notified just as a permanent depot would have been and beat engineers dispatched from it in patrol cars parked outside. Within a few years the mobile depots were phased out as the network of permanent depots became better established.

With the rapid spread of the SP Alarm System, Iida felt the need for a company name that would more accurately mirror the corporation's expanded activities. SECOM, standing for Security Communication, was the term Iida chose to reflect both the security and the communication aspects of the business.

> This term signifies *anzen joho kagaku*, literally, 'safety information science'. Contemporary society demands new security systems which are a marriage of man and science. The establishment of such systems requires a synthesis of related sciences

including mechanical engineering, disaster prevention engineering, electronics, computer technology, mathematics, telecommunications, optical science, acoustical science, psychology, and medical science. The new scientific technological field comprised of diverse scientific knowledge is what the term SECOM or Security Communication represents.

The term was first used publicly in an advertisement carried by four major newspapers with national circulation in February 1973. The advertisement introduced the concept of a SECOM society, a peaceful society where security is assured. At about this time, too, Iida first suggested the possibility of a company name change to his staff. Successful entry of the corporation into the new era, Iida believed, required determination, energy, and creativity. Iida felt that these characteristics were prevalent in his corporation but he knew that it would take several years for his staff to fully appreciate that the corporation was accomplishing its mission. To penetrate the new world Iida wanted to have a new banner. He felt that the emotional attachment to and nostalgia surrounding the name Nihon Keibi Hosho were too great. Changing the name to SECOM would force employees into a new way of thinking while still retaining their belief in the company's social mission.

The idea of a name change met with loud opposition, especially from those directly involved in security guard activities. Employee resistance to the proposed name change was symbolic of their resistance to the rapid ideological and technological changes that were taking place within the company. Under the circumstances Iida thought that addressing the name change issue within the context of resistance to change would only intensify the opposition and muddy the issue. He considered the solution of this problem as a fundamental managerial challenge. This was an opportunity to instil in the minds of the employees the importance of building and maintaining an innovative organization where change would be the norm rather than the exception and an opportunity rather than a threat.

The name Nihon Keibi Hosho remained in use where it already existed but Iida and his staff agreed that the term SECOM would be incorporated into the names of new branches, new affiliated and associated companies, and new products. The following year when the company opened its employee training centre in Yokohama the name SECOM was visible everywhere: on towels, *yukata* or 'robes', electronic equipment, cutlery, napkins, ashtrays, and meal tickets. Within two years all company vehicles and mobile depots carried the name SECOM.

Gradually, SECOM came to be used interchangeably with Nihon Keibi Hosho but it did not become the official corporate name until 1983. Iida's intention when he chose the term SECOM was eventually to expand his security corporation into what he terms a 'social system industry' with security, understood in its broadest sense, as the foundation. Aligned with corporate values, Iida's vision for the company reflected in the new name created a sense of purpose for the employees. The name change was accompanied by employee recognition that Iida's vision was worthwhile and attainable, although slightly beyond their reach at that time.

LEGAL CONTROLS

Increased public awareness of Nihon Keibi Hosho's services through media coverage and the company's expanding business frontiers and catchment areas triggered the establishment of numerous security guard businesses throughout Japan. Prior to 1972 there were no laws regulating security guards. Anyone could found a company or be employed by such an establishment. By the end of 1971 the number of security guard establishments in Japan had grown to 448 although the majority of them were small enterprises. Their employees totalled 34,491.[4]

The involvement of some of these firms in scandals became the target of public outcry. The first six months of 1971 saw three such scandals: the hiring and use of security guards by the mayor of Nakaminato city in Ibaragi prefecture; the involvement of the Narita Airport Security Corporation in the dispute over land procurement for the proposed Narita International Airport; and the involvement of security guards in the railroading through of a shareholders' meeting of Chisso Corporation, the company accused of the mercury poisoning of Minamata Bay.

Nakaminato City Hall scandal

On the morning of 12 January 1971 several muscular young men, dressed in a manner resembling gangsters with their black suits and sunglasses, entered Nakaminato City Hall. After removing posters hung by union members protesting the mayor's autocratic rule they wandered through the building intimidating workers. These men had been hired by Mayor Yohei Usui to quell the growing unrest among members of the city hall union. They belonged to Tokubetsu Boei Hosho, a Tokyo-based security guard company with strong right-wing leanings. Anticipating a confrontation, the police arrived on the scene but did little more than observe the situation.

Yohei Usui had been in politics for sixteen years, eight in prefectural politics and eight as mayor of Nakaminato. Initially an employee of the Fishery Agency, he had gone on to become managing director of the Nakaminato Fishing Co-operative. Nakaminato, with a population of 33,000, was entirely dependent on its fishing and fish processing and distributing industries. It was a hierarchical society where some twenty families had, since the middle ages, controlled municipal politics and determined who would be the city's prefectural representative. Throughout his political career Usui had only once been elected. On the other occasions there had been no opponents and he had gained office by acclamation.

Beginning in 1968, Usui initiated a series of rationalization programmes to restructure the city's administration, abolishing upper management and introducing an ability-based evaluation system where a traditional seniority system had prevailed. He sought to improve the productivity of public sector workers, emphasizing the need for an entrepreneurial outlook at city hall where workers previously tended to coast. In the spring of 1970 Usui had ordered a moratorium on wage increments for eight employees. When the head of the labour union and his secretary protested, Usui fired them. By the end of the year Usui had forced nineteen city employees to take leaves of absence. In response, union members, 60 per cent of all city employees, expressed their displeasure by appearing at work in sweatbands bearing anti-administration slogans. Section managers could not control the workers. Frustrated, the mayor called in reinforcements in the form of 'body guards', further angering city employees.

Five guards stayed in the room next to Usui's office, an area off limits to other employees. The guards' activities were open to speculation but they appeared to be practising karate and boxing. The remaining fifteen guards stayed in a nearby hotel and received token new employee 'education' for two hours each day. They passed the rest of the time strolling on the beach and resting. The situation worsened when city employees learned that even in their probationary period the guards were receiving higher than average pay.

The mayor threatened to increase the number of guards to fifty, giving the city's finance officer cause for concern. He had not been notified initially of the addition of twenty persons to the payroll and now he wondered how he could meet the additional salaries, short of redirecting funds from major construction projects. The situation was complicated by the fact that the mayor was breaking the law in placing on the city payroll personnel in the employ of another corporation, in this case Tokubetsu Boei Hosho.

The prefectural government became involved at this point in an attempt to resolve the crisis. Reporters wishing to interview the guards at their hotel were verbally and physically threatened. Two socialist members of the National Diet came to Nakaminato to register their support for the workers. Recognizing the seriousness of the situation the mayor attempted to fire the twenty bodyguards but met with resistance on their part. The mayor eventually found a way to dismiss the guards at a considerable cost to Nakaminato tax payers.[5]

Narita Airport scandal

The Narita Airport incident occurred at the end of February 1971. It marked the culmination of dissention that had been building for several years. By the early sixties Haneda International Airport in Tokyo had reached the limits of its capability to handle air traffic. The decision to build a new international airport was made in 1962 and four years later farmland in the Narita environs 70 kilometres northeast of Tokyo was selected as the site, to be opened in the spring of 1972.

The Narita Airport Construction Corporation was established and entrusted with land purchase. Three hundred and twenty-five households within the airport precincts and 412 households within the noise pollution periphery were to be relocated. The Chiba prefectural government and the Narita Airport Construction Corporation made available land for relocation. In addition, compensation was to be paid to each family. Of the 325 families living on what was to become the airport site, 270 (83 per cent) were agreeable to the settlement and by the spring of 1969, 250 households had moved to new accommodation. Many of them had abandoned farming for new occupations such as shop ownership, construction work, and taxi-driving.

Others had found employment as guards with the Narita Airport Security Corporation. This corporation was established in February 1968 by seventy-three relocated farmers using their compensation money. It was their intention to be involved in every aspect of airport security and to provide a security guard service to business firms in the Narita area. The staff numbered 200 among whom twenty were part-owners of the corporation. The remaining employees were drawn from former Self-Defence Force members, police, and farmers.

Objection to airport construction was strong among the fifty-five households resisting relocation from the airport site and nearly all those within the noise pollution periphery. Attachment to the land, distrust of the government, and a certain innate stubbornness were at the heart of the dissention.

The conflict became more complex when, in 1968, the Kyokusa Boryoku Shudan or 'extreme leftist mob', believing that the new airport was intended to serve as a military base from which Japan would feed the American war effort in Vietnam and would become the base from which Japanese imperialism could be spread overseas, threw their support behind the farmers and labourers. Together they planned their strategy and, by the time of the first major confrontation in February 1971, on the land marked for appropriation they had constructed a headquarters, observation towers, six forts, communication stations and a twenty-bed field hospital staffed by fifty students including medical students. They had armed themselves with gasoline bombs, bamboo spears, and wooden shields.

The prefectural government, arguing that the public good must take precedence over private interests and recognizing that further attempts at negotiation would only delay the airport opening, decided to take the land by force. Employees of the Narita Airport Construction Corporation, 200 guards from the Narita Airport Security Corporation, and several hundred riot police launched an attack on the protesters who by this time had swelled their ranks by the addition of fifty junior high school students whose parents or adult friends were involved.

The protesters surrounded themselves with old tires and wood and set the barricade ablaze. Despite the fact that the riot police and guardsmen had orders not to inflict physical injury, there were casualties on both sides during the week the confrontation continued as former friends and neighbours, now protesters and security guards, fiercely battled against each other.[6]

The conflict intensified and resulted in the deaths of three riot police on 16 September 1971 when 15,000 protestors fought police and security guards. The ongoing dispute slowed down but did not halt the airport construction. On 20 May 1978 the airport was finally opened, six years behind schedule. The dispute has still not been satisfactorily resolved and more than ten years after opening the airport remains under heavy security.[7] The Narita confrontation brought into question at the national level the function and legal standing of security firms whose guards wore uniforms resembling those of the police and who used nightsticks as weapons.

The Chisso scandal

The third incident in which security guard involvement received negative publicity was the Chisso Corporation annual shareholders' meeting of 26 May 1971. The Chisso Corporation became the target of violent demonstrations after corporate irresponsibility led to mercury pollution in Minamata Bay with devastating consequences for the health of area residents once the mercury made its way into the food chain.[8] The corporation's responsibility for the pollution was widely acknowledged and representatives of those suffering from the debilitating and often fatal Minamata disease were expected to put the question of corporate social responsibility before the shareholders' meeting.

To assure a good seat at the meeting, 700 people including representatives of those stricken with Minamata disease, company employees, and *sokaiya* (literally 'general meeting fixers' or professional manipulators of shareholders' meetings), spent the night before the meeting camped outside the hall where it was to be held. They and others had each purchased one Chisso Corporation share to entitle them to a voice at the meeting.

Sokaiya numbering several thousand are usually associated with the *yakuza* or 'Japanese Mafia'.[9] A traditional *yakuza* activity is to blackmail companies by threatening to raise embarrassing questions at shareholders' meetings. Companies often pay generously to prevent this. On the other hand, when a company anticipates awkward questions from legitimate meeting participants, *sokaiya* may be hired to shout them down. The latter was the plan of the Chisso Corporation.

When the meeting convened the 2,000 seat hall was filled to capacity. Four hundred security guards were at the site to maintain order, sixty of whom the Chisso Corporation had hired from Tokubetsu Keibi Hosho, the same company that had supplied guards to the mayor of Nakaminato city. The *sokaiya* took control of the meeting and within fifteen minutes it was over. During and after the meeting the security guards and the *sokaiya* joined forces to control the enraged victims.

The use of *sokaiya* by large corporations to deal with dissenting voices at shareholders' meetings was common. Many companies were under attack for corporate irresponsibility ranging from environmental pollution to heavy involvement in the production of military weapons to be used in the Vietnam War. To avoid embarrassing questions shareholders' meetings would be terminated as quickly as possible. For example, in 1971 the annual meeting of the steel producer Nippon Kokan lasted only fifteen minutes while that of Mitsubishi Heavy Industries was finished in twenty-five minutes.[10]

The way in which certain Japanese corporate executives dealt with issues related to social responsibility at shareholders' meetings was indicative of a widespread lack of a social conscience. The use of *sokaiya* and security guards for corporate advantage in particular became a nationwide scandal.

National security guard industry law

Although the three aforementioned incidents were the most publicized, they were not the only situations in which security guards were criticized for abuse of force. Security guards also came under fire for their involvement in the student–administration/faculty confrontations on university campuses throughout the country. The reputation of the entire security guard industry was further tarnished by the apparent link between some companies and the *boryoku-dan* or *yakuza*.

Today Japan's National Police Agency estimates that the country has 80,000 gangsters but in the early seventies their numbers were closer to 180,000. The public outcry against *boryuku-dan* involvement in security guard corporations was strong. As of March 1971, seventy-seven senior executives within Japan's security industry had criminal records. Of these, twenty were company presidents. Moreover, some 100 guards had criminal records. In 1970 alone sixty-five guards were convicted of criminal offences including theft, assault, rape, hit and run, and fraud.[11] Sensitized by his previous experience with internal theft, Iida continued to be careful about screening prospective employees, rejecting any with suspected *boryuku-dan* connections.

He considered which types of security needs his company would fill and which would be avoided. Distinguishing between acceptable and unacceptable contracts was an obstacle faced by security firms around the world. Adopting the position of Ligue Internationale des Sociétés de Surveillance, Iida's decision was to maintain neutrality. Solutions to the types of problems discussed above, he felt, lay in negotiation between the concerned parties, not in security war games. Despite Iida's convictions, Nihon Keibi Hosho's reputation suffered as a result of the unfavourable publicity attracted by less savoury industry members.

In response to public pressure the Liberal Democratic Party asked the National Police Agency to submit a proposal regarding legal controls that should apply to security firms. While the police had generally welcomed the initial emergence of private security firms, feelings of jealousy and annoyance were now beginning to arise in some quarters. It was under these conditions that the National Police Agency began preparation of its draft of a security law which was presented to a meeting of over fifty Tokyo security firms in August 1971.

In the ensuing debates, consideration was given to the following concerns, amongst others: 1) whether or not security guard companies should be licensed; 2) if they were licensed and malpractice on the part of a firm was identified, how much authority the police should have to investigate corporate activity to determine if the licence should be revoked and for how long; 3) what qualifications guards should be required to meet and whether security firms or the police should decide who met these qualifications; 4) whether or not potential guards should be given basic training and if so what the content and duration of this training should be; 5) the extent to which guard uniforms and patrol cars had to be readily distinguishable from those of the police; 6) the extent to which security guard corporations should provide compensation to clients when the security system failed to give adequate protection; 7) whether or not security companies should be legally bound to honour the privacy of their clients; and 8) whether or not the law should specify the establishment of a nationwide security guard association.

The National Police Agency advocated a minimum of regulation governing the security industry. Their preference was for the industry to be self-regulating, a reflection of growing societal concern for what was perceived as excessive government intervention into financial and industrial activities. When the Security Guard Industry Act was introduced in the Diet (Parliament) the Socialist and Communist parties raised strong objections. The purpose of the law, they argued, was to nurture the growth of a high quality, effective security industry which would eventually develop into a second police force to be used by the government to stamp out labour unrest.

Despite the objections of the opposition parties, the Security Guard Industry Law was enacted in October 1972 and imposed minimum controls on the industry. Licences were not required but the establishment of new firms had to be reported. Details of uniforms and protective equipment had to be registered with the police along with the geographic areas of operation of the business. Personal information relating to each guard employed by the company, including his photograph and indication of his completion of the necessary training, had to be submitted. Each time a contract was signed, the client's name, duration of the contract, location, type of service to be provided, and the number of guards involved had to be given to police.

The law indicated the content, method, and duration of in-house education for security guards. Guards new to the business were required to undergo in excess of twenty hours training prior to their first assignment. Those who had previous experience in excess of one year with other security guard firms or with the police force were required to have five hours of training. All guards had to be given ten hours additional training each year. Through textbooks, lectures, and hands-on experience guards had to become familiar with the fundamental rules of security activities, professional ethics, laws relating to security work, emergency procedures, and the handling of protective gear. In 1972, Zenkoku Keibigyo Kyokai Rengo-kai (The Federation of Associations of Security Industries) was formed in order to improve and standardize security services throughout the country. One of the federation's first tasks was the education, through seminars, of security firms with respect to the implementation of the new law.[12]

Security firms continued to increase in number and became part of the everyday life of Japanese society. By 1975 they numbered 1,682 with a total of 71,333 employees. Their operations brought them into frequent contact with criminal activities and they supported the work of the police in the maintenance of public safety in general. In 1975, for example, security firms assisted in the arrest of 5,905 criminals, saved 287 lives, and were responsible for the early detection of 43 fires.[13]

The sudden upsurge in security firms in the mid-seventies reflected the rapid institutional adoption of security services. Between the summer of 1974 and the spring of 1975 there was a series of bombings targeted at major corporations. The first occurred at noon on 30 August 1974 at Mitsubishi Heavy Industries' headquarters in Tokyo. The explosion killed eight and injured 380. Ten other bombings followed, the last one on 4 May 1975. The targets were Mitsui and Company Ltd, the Research Institute of Teijin Corporation, Taisei Construction, Kashima Construction, Hazamagumi Construction's head office, plant, and work site, the Korean Industry Economic Research Institute, and Oriental Metal. With the exception of Oriental

Metal, which was located in Amagasaki city in Hyogo prefecture, all were in the greater Tokyo area.

The bombings were carried out by three small groups: 'Wolf', 'Fangs of the Earth', and 'Scorpion', all members of the East Asian Anti-Japanese Military Front. Group members, particularly those belonging to Wolf, had earlier been associated with nationwide university unrest. They were driven by their belief in the right of the labouring class to revolt against the giant and/or multinational corporations which were exploiting it.[14] The groups appeared to attack indiscriminately, causing widespread fear. In defence, companies tightened their internal security and hired private security firms.

Despite the heightened awareness of the security industry, many security firms failed to meet the requirements laid out in the Security Guard Industry Law. Penalties were rarely applied against those firms who chose to ignore the law, which proved to be too vague and lacked teeth. Consequently, the quality of security service degenerated. The excessive competition that resulted from the rapid increase in security firms brought about severe reductions in fees charged for their services as companies tried to undercut each other. The greatest cuts were in bids for government contracts which were automatically awarded to the lowest bidder. Companies were often willing to take a loss in order to secure these contracts in the hope of winning credibility in the eyes of the private sector. Occasionally, the fees charged to the government were not even sufficient to cover the guards' salaries, transportation costs, and uniforms, not to mention training and education.

Many firms went bankrupt. Pacific Keibi Hosho, in existence for ten years, is one example. A respected corporation in the public's eye, its customers included the Supreme Court, the American and Korean embassies, and several major airports. Misguided foreign investment in Guam, the large capital investment required for entry into the field of machine security, conflict between the company's two senior executives, and the accumulating debt arising from excessive underbidding on government contracts all eventually led to the company's demise. When bankruptcy was declared the company's president vanished, embarrassing the security industry as a whole.[15]

To remain competitive, firms cut corners at every opportunity. Small, financially weak companies resorted to hiring inexperienced part-time personnel whom they could not afford to train properly. In many cases there were no funds to hire additional guards needed to meet the increased work load and existing guards were overworked, often in violation of the Labour Standard Law. While the problem was recognized by the newly formed Federation of Associations of

Security Industries, to which approximately half of the security firms belonged, and by the Ministry of Labour, neither had the power to curtail the situation. A high employee turnover rate brought about by poor working conditions plagued the security service industry. Morale was low and inappropriate personnel were frequently hired. In 1975, for example, 139 thefts were committed by on-duty guards.[16]

Some corporations thought that the introduction of alarms and other security equipment would reduce costs significantly. Technological innovation, however, was haphazard, was not carefully integrated into corporate planning and as a result undermined the reliability of the security service. Client complaints abounded. Frequent false alarms triggered by malfunctioning equipment took their toll on the police as well.

The law remained unchanged for ten years. Amendments introduced in 1982 received the support of the Communist Party as well as the Liberal Democratic Party. Enacted in 1983, these amendments rendered the law more specific and more readily enforceable with penalties for violation including imprisonment. By this time the number of firms had increased to 3,546 with a total of 133,946 employees.[17]

Under the new amendments security firms had to be licensed and the qualifications for both owners and guards became more stringent. Guards were required to undergo educational training conducted by licensed instructors who themselves had completed National Public Safety Commission Seminars. With regard to electronic security systems, details of their operation had to be filed with local branches of the Public Safety Commission. System owners would be licensed by this body and would be required to complete its training seminars. The Public Safety Commission was authorized to investigate the activities of any security corporation and, when necessary, to prosecute.

7 Cultivation of human resources

STOCK EXCHANGE LISTING

With the rapid growth brought about by the popularity of the SP Alarm System Nihon Keibi Hosho's total sales exceeded ten billion yen in 1973. The character of the company was changing as it capitalized on the computer network information system. On 21 June 1974 a showroom was opened to the general public in the company's main office in Akasaka, Tokyo where components of the newly developed alarm systems could be viewed. To support continued expansion into new frontiers, Nihon Keibi Hosho needed to strengthen and expand its financial resources. Iida felt that the time had come to go public.

In addition to providing the company with a broader and more secure financial base, Iida saw three other advantages to the move. Stock Exchange listing would bring societal recognition of the company as a leader in its field. The resultant employee pride and heightened awareness of social responsibility would bring about improvement in service. Greater public acceptance would also facilitate employee recruitment.

While all corporations come under close government scrutiny before they can be listed on the Stock Exchange, for Nihon Keibi Hosho the examination was more stringent than usual because it was the first security firm to apply for listing. The company's quality of service, the effectiveness of its organizational structure, its administrative network, financial constitution and managerial philosophy were all examined within the context of the distinctiveness of the service provided. The company fared well under inspection.

On 24 June 1974, twelve years after establishment, Nihon Keibi Hosho was registered on the second section of the Tokyo Stock Exchange. The nature of this particular service industry was so new that there was no ready-made industrial classification into which it would fit, thus it was relegated to the end of the miscellaneous section of the Stock Exchange. In his speech to his employees marking the occasion Iida stressed corporate responsibility and innovative attitude:

> Just as a child grows into adulthood, so listing on the Stock Exchange marks Nihon Keibi Hosho's entry into corporate adulthood. . . On this occasion we must reaffirm our confidence in our future development and at the same time give increased recognition to our responsibility towards society at large. As we mature from childhood into adulthood, we must not lose the youthful inspiration and intellectual curiosity that has propelled us in the past to seek new knowledge and explore new frontiers. It is my hope that our corporation will maintain these strengths and will continue to grow. Together we can make this happen.

In accordance with Japanese custom, Iida wished to send a congratulatory present to the family of each employee to mark the change in the company status. Respecting individual needs and tastes he decided on a gift certificate redeemable at a number of different stores. This was accompanied by a letter thanking each employee and his family for supporting the corporation and making possible its growth.

Iida used the occasion of Stock Exchange listing to change the end of the corporate fiscal year from 31 December to 30 November. Separating the fiscal year end from the calender year end allowed the staff time to complete corporate activities before preparations began for New Year, Japan's most widely celebrated holiday. The 1974 fiscal year had only eleven months. Even so total sales reached nearly fifteen billion yen, up 50 per cent over the previous year. Income from the SP Alarm System accounted for nearly 70 per cent of total sales compared with 50 per cent the previous year. The fiscal year end would remain 30 November until 1990 when, in order to bring the company in line with government and financial institutions and the majority of private companies, Iida moved it to 31 March.

With continued growth and improved performance, in May 1978, four years after the initial Stock Exchange listing, Nihon Keibi Hosho moved from the second to the prestigious first section of the Tokyo Stock Exchange and in 1986 was listed in the first section of the Osaka Stock Exchange. In 1978 the company moved its main office to the Nomura Building in the Shinjuku district of Tokyo where it is still located. This was the eighth move for the central business office. The first six moves occurred in the first eight years after the company's establishment. Each move was to larger premises reflecting the

7 Control Centre, Tokyo, 1978.

company's continuous growth and corresponding adjustment.

Nihon Keibi Hosho was expanding quickly. Every year since its establishment sales and profits had increased. In 1977 total sales were 28.6 billion yen and net profit was 2.1 billion yen, in each case an increase of approximately 20 per cent over the previous year. In mid-1978 *Nihon Keizai Shimbun*, Japan's most authoritative financial newspaper, ranked 1279 corporations listed on the first and second sections of the Tokyo Stock Exchange according to six criteria: initiative, productivity, size, effective use of corporate assets, degree of social contribution, and future prospects.

The newspaper pointed out that among the top-ranking corporations the product and service market share was substantial, the equity ratio was high, the entrepreneurial spirit was clearly evident in such factors as product innovation, and the internal/external adjustment to the changing environment, including a rising yen and a domestic business slump, was appropriate. The top-ranked corporation was KDD (Kokusai Denshin Denwa), an international telecommunication company. It was followed by Toyota Motors with Matsushita Electric Trading in third place. The latter is a medium-sized trading company which has exclusive charge of exports for the Matsushita Electric

Industrial group. Nihon Keibi Hosho ranked fourth. With this recognition Iida resolved that the company would continue to grow as an innovative corporation and would contribute to the healthy development of Japan's security industry.

EMPLOYEE TRAINING

Living in a country with a scarcity of natural resources Iida was acutely aware that competitive advantage must be built largely on human resource development. Employee education, he believed, was an effective corporate investment. Especially when the business is entirely new to society, growth and development depend on the entrepreneur's success in nurturing employees who understand the corporate ideology.

For Iida, management is a vehicle for self-expression and, he believes, it is also a channel for the employees' self-actualization. In Japan, approximately thirty years of a male university graduate's life revolves around his workplace. Thus, it is essential to create the appropriate working environment.

Our corporation was challenged to develop a new service: the provision of security. In order to actualize our mission, the organization as a whole continuously seeks improvements in our service. We are committed to meeting new challenges. This innovativeness and forward looking propensity is a distinctive feature of our corporate personality.

The development of an appropriate working environment is important not in regard to strengthening employee loyalty towards the corporation but rather, because, through the understanding of the purpose of one's work and how it contributes to society, one develops a feeling of self-worth and satisfaction.

Employee utterances like 'I exist for the corporation; I give my all to the organization' do not please me. The individual must place primary importance on himself. One has to grow, has to satisfy one's own needs first. Ideally the outcome of such effort will be linked to the progress of the organization. I encourage my new employees to maintain the optimistic, unfettered attitudes of youth.

At the same time, I do not want my employees to be individuals who can relate only to the corporate environment. I want them to associate with a wide variety of people and broaden their horizons. Yet, employees should share the fundamental values of our corporate culture which is oriented towards an eternal search for the ideal state.

Corporate culture, Iida feels, should permeate from the top down. With the intention of cultivating a group of managers who shared the same vision, Iida introduced the first formal education seminars for branch and main office managers and guard supervisors in the autumn of 1968. Held at Yomiuri Land, a recreational facility near Tokyo, the purpose of the seminars was to clarify to each attendee his place and role in the corporation's progress. Lectures by internal and external speakers, case study discussion, and self-analysis and evaluation through sensitivity training were the primary educational methods employed.

The permeation of the corporate culture, Iida realized, could be ensured only by his constant reiteration of it with groups of employees at all levels. This was especially true in the early years of the company's development when so many of his staff had come from other corporations, each one bringing with him values and attitudes shaped in part by his previous employer. At that time Iida's primary approach to employee education was to talk informally with individuals and small groups, often after hours. With the formal multi-level staff development sessions introduced in the late sixties, from time to time, about thirty employees from all over Japan would get together for a few days at Yomiuri Land. At the conclusion of the structured programme Iida and the men would gather to exchange ideas informally. On these occasions Iida would talk about the corporate mission and how a given corporate strategy was emerging in accordance with this mission.

His objective was to explain strategic direction so that individual employees could get a clear picture of the way in which their particular jobs related to overall corporate strategy. Repeated interpretation and reinforcement was necessary, Iida realized, especially for those performing tasks not seemingly meaningful in and of themselves. In Japan, because of society's emphasis on cooperative effort, employees tend to recognize their contribution to the overall running of the corporation but, nevertheless, Iida knew that conscious effort was required to boost morale.

These sessions helped employees to see the corporation as a whole and to understand their role in its development. Yuji Sato was a security guard working out of the company's Nagoya branch when he attended his first seminar towards the end of 1968. Lacking a firm commitment to his work and dissatisfied with some aspects of corporate life – physical labour, hard work, long and irregular hours – he had been thinking about resigning. Through attendance at a staff development session he was able to view his work differently and see a social significance in his job. From then on he became highly motivated and committed.

Takushi Sakaguchi was another employee on whom the training session had a profound impact. He had been experiencing feelings of emptiness and alienation on the job. Through exposure to corporate philosophy, planning, and organizational structure he was able firmly and positively to relate himself to the company. Also, through the association with people close to his own age working in other branches he sensed that he was a part of the collective energy of a dynamic whole. His feelings of alienation were dispelled.

Sakaguchi was one of the many employees with whom Iida had direct contact in the early years of the company's development and who was inspired and motivated by this. Over the years Iida's charismatic personality has been a source of energy for those who have come directly under his influence. As the corporation grows, however, there are more and more employees who have never had personal contact with Iida, having only second-hand knowledge of his views and corporate philosophy.

The seminars at Yomiuri Land also included practical training. For example, in preparation for the introduction a 400 MHZ shortwave frequency for commercial use, technical and legal instruction was provided by Nihon Keibi Hosho specialists as well as by external experts in the field. Iida and Toda as well as employees from all over Japan already licenced as shortwave radio operators, control centre personnel, and twenty-two guards working closely with the SP Alarm were in attendance for the first two-day seminar. Lectures dealing with new technology came to be an integral part of employee training as technological advances were incorporated into Nihon Keibi Hosho's security system.

Previously, verbal communication between on-site guards and the control centre depended on public telephones and walkie-talkies since private security firms were not granted licences for shortwave communication. The growing importance and visibility of security firms throughout Japan resulted in the government decision to allow them access to shortwave communication as of August 1970.

In addition to Yomiuri Land Iida also held employee training seminars on the S.S. Sakura, a 13,000 ton ship built in 1962 and used initially to display Japanese goods at foreign ports in some eighty countries. The first Sakura seminar began on 20 September 1973 and lasted eleven days while the ship cruised from Yokohama to Hong Kong and back. The one hundred participants were being trained as instructors for various groups of forthcoming new employees. Altogether about 900 employees participated in these floating seminars before the company opened its own training centre in 1974.

Employees from all ranks participated in these seminars which ranged from the practical to the conceptual, according to employee level and need. In addition to the external 'experts' Iida would bring in for the occasion, he frequently lectured himself on topics he felt to be of significance.[1] By this time Nihon Keibi Hosho had developed a sizeable knowledge bank relating to security matters and employee education. Many corporations provided in-house security. Some of the larger firms had created subsidiaries staffed largely by their former employees to handle their security needs and to provide employment for their retirees. Gradually they began to seek out Nihon Keibi Hosho to help train their own guards. Iida considered this a viable way to instil security awareness.

In addition to the provision and exchange of security information on the domestic front, Iida became involved on the international level through Ligue Internationale des Sociétés de Surveillance. At the Ligue's 1969 conference held in Australia it was decided that, in order to groom future top executives in the security field, young executives from member firms would visit security companies in other countries to exchange and broaden knowledge. As a result, Christopher Bottomly from the Australian member firm, after spending ten months in North America and Europe, came to Nihon Keibi Hosho in the autumn of 1969. He spent three weeks under the wing of the company's educational training section visiting company branches and their clients throughout Japan. He also spent time at the Tokyo control centre and observed executive training sessions. At this time training sessions were held in rented facilities but with Nihon Keibi Hosho's rapidly expanding work force and management body Iida saw the need for a company-owned training centre.

THE HUMAN DEVELOPMENT CENTRE

In September 1974, the Chuo Kenshujo or Central Training Institute was opened in Aobadai, Yokohama city and training could now take place as frequently as required in nearly ideal surroundings. Encircled by a garden, the building had three storeys plus a basement. On the ground floor was a large lobby, restaurant, cafeteria, office space for instructors, VIP lounge, and the administrative offices. On the middle

8 Iida (centre) and Hashimoto (far right standing) with SECOM's rugby team at its inauguration, 1985.

floor was a large lecture hall, seminar rooms, simulation facilities, and an audiovisual centre which included a studio for video production. The top floor had accommodation for sixty-two people and library facilities. A large communal bath and a sauna were located in the basement. When the Institute was not being used for SECOM training sessions it was available for rent to other institutions.

As employees increased in number and company activities diversified, Iida felt that employee understanding of corporate values, managerial attitudes, decision making skills, and specialized product and service knowledge was lagging. He realized that his employees were at the cutting edge of activity and that it was essential to provide them with the opportunity to grasp more fully the overall workings of Nihon Keibi Hosho and their importance to corporate success. To this end the Human Development Centre was established in Gotenba in Shizuoka prefecture, close to Mount Fuji. Opened in June 1981 the enlarged facility includes lecture halls, seminar rooms, simulation facilities, bedrooms, dining rooms, and recreational facilities. The Aobadai Training Institute has been converted to a dormitory for single male employees working in the Tokyo area.

It has been pointed out that the growth of the security industry and the expanding use of employee indoctrination techniques and sophisticated alarm and monitoring techniques may create the impression that society is moving in the direction of the totalitarian state described by George Orwell in his classic *Nineteen Eighty-Four*.[2] Considering the speed with which the microelectronic revolution is advancing it is not surprising that there are some who feel that the day is coming when human life will be ruled by the silicon chip. They overlook the fact that the sensors, cameras, and monitors are simply extensions of the eyes and ears of security personnel and as such are only as good as the people who act in response to their signals. People are the heart of security service. It was the 'human touch' that made Nihon Keibi Hosho distinctive. In the future the company would develop into a high technology-based communications industry but its security personnel would remain the cornerstone of the corporate foundation.

While Iida placed extreme importance on technological research and development in relation to traditional security service, he was also contemplating expansion of the business frontier in other directions. The opening of the Gotenba Human Development Centre coincided with Iida disclosing to the staff his intention to establish a social system industry with security at its core. He explained to his employees that Nihon Keibi Hosho's concentration had been on theft and fire security. His objective now was to expand the firm's activities using the company's already established communication network system, the largest private system in the country. He envisaged the provision of an emergency medical alert system, business database services, reservation services, and total building and facilities maintenance.

Offered by a variety of corporations, these services existed independently of one another. Iida planned to draw them together for increased customer convenience at reduced cost. For only a slight additional expense customers would have access to a wide range of services in addition to their basic security package. The strength of his proposed social system industry, Iida reminded his staff, was the company's integrated network system infrastructure comprised of the computer, telecommunication circuits, and the human activities for emergency maintenance and operation services. To convey this idea to his employees, Iida likened this system to the head, the nervous system, and the muscles of a human being.

In the Gotenba Human Development Centre training sessions are provided for all employees according to position and skill requirements. New employees (just out of school) attend a series of lectures given by personnel from different levels of the corporation from Iida himself down to experienced workers. Topics include corporate ideology, market standing, future directions, professional ethics, and practical instruction on the accomplishment of work routines. After the first few days the lectures are interspersed with seminars and discussions. New employees are assigned to on-the-job training at designated work sites in the second week. Six months later they return to the Centre for a few days of follow-up lectures and seminars.

Employees coming to SECOM from other corporations undergo a similar training programme but the emphasis is on the smooth transition from one company to the other. Ongoing education is also provided at the Human Development Centre. Employees return from time to time throughout their careers for skill acquisition and upgrading. For many, this education is preparatory to writing a variety of in-house examinations upon which promotion decisions at the pre-executive level are based. The Centre is in constant use. In 1987, for example, four thousand employees received some form of formal education there.

To keep pace with the ongoing increase in employee numbers there is a constant need for additional training centres. The Aso Centre in Kumamoto prefecture opened in August 1990 to facilitate SECOM's one thousand employees in the Kyushu district. In addition to functioning as a training centre the Aso facility serves as a holiday resort for employees and their families.

Although Iida personally does not favour the idea of employees spending their private holiday time with each other in facilities provided by the corporation, as is common in the Japanese corporate world, he acquiesced to employee pressure for such facilities in the late sixties. In 1969 a recreational centre with accommodation for twenty-two people was established in Nagaoka Hot Springs on Izu Peninsula. This was followed by centres in Karuizawa and on Awaji Island.

Despite the fact that they were built in response to employee demand, these facilities are not well used and Iida's recent approach has been to incorporate their function into the training centres. Plans are under way for the construction of five training centres in other parts of Japan with construction having already begun on one in Nahari City in Mie prefecture to accommodate employees in the Chukyo District of central Japan. The total investment for the five centres is expected to be five billion yen.

STUDY SCHOLARSHIPS

A month after the opening of the Central Training Institute Iida introduced a scholarship plan to promote domestic and overseas study by Nihon Keibi Hosho's young employees. Domestic scholarships are available to those high school graduates under 21 years of age who have been with the company a minimum of one year. These are four-year scholarships for study at a domestic university of the employee's choice. These scholarships are geared to individuals who would have liked to have gone on to university after high school graduation but were unable to do so. Successful scholarship applicants must pass an internal screening process and be accepted by the university.

The scholarship covers such expenses as entrance, enrolment and tuition fees, and commuting costs as well as dormitory rent and two meals a day. Recipients are obligated to participate in on-the-job company training at the appropriate branch office one or two days a week depending on their study timetable during the school term and full-time during vacations. They receive their regular salary for these days. During the four years of study employees are affiliated with the personnel department of the company and continue to acquire time-based seniority. Despite the generous terms of the domestic scholarship as of 1991 no employees had taken advantage of it. The international scholarship on the other hand, has had wide appeal.

Iida felt that developing an international perspective on both the individual and corporate levels was essential for corporate and personal growth in the coming age of internationalization. It was for this purpose that he introduced the foreign scholarship programme. It pays for travel expenses, tuition, books, and living expenses up to a specified limit. Candidates must be single, under twenty-seven years of age, have been with the company over a year, meet the prescribed English language fluency requirements, and be recommended by their superiors. Those who meet these requirements must write an essay in both Japanese and English explaining why they wish to study abroad and outlining their proposed study plan. Next they are interviewed by Iida and the personnel manager. Finally they must be accepted by the recipient institution.

In 1975, the first year the programme was in place, two employees were successful in their application for scholarships. Yohichiro Matsuo, a 1974 graduate of Nippon University's Department of Commerce, was employed in the company's accounting department. He selected a small institution in California at which to study English and international economics for two years. The other scholarship recipient was Susumu Hirata. He graduated in 1973 from Hirosaki University Faculty of Science. After working for another company for a year he joined Nihon Keibi Hosho in 1974 and was involved in fire extinguishing research. He studied chemical engineering at Oregon State University after first completing an English as a Second Language course in Seattle.

Since the programme's inception there have been about 110 successful candidates. All have chosen to study at universities in the United States including Boston University, University of California, Carnegie Mellon University, Massachusetts Institute of Technology, New York University, Oklahoma State University, University of Pennsylvania, University of Southern California, and University of Washington. As well as formally pursuing degrees in the sciences, economics, business, and language, informally scholarship winners improve their communication skills and learn a great deal about American culture. Some take the opportunity to expand their business network. Says Hiroyuki Kirishima who received an MBA at Boston College in 1987 after two years of study:

> During my stay in Boston I divided my time equally between Americans and Japanese. Because I was sent by my corporation I thought it advisable to take the opportunity to contact other Japanese sent by their companies studying at my college and also at MIT and Boston University. Expanding my business connections in this way would be of lifelong benefit once I was back in Japan. I made many American friends too both on campus and off.

During the time they are abroad scholarship recipients write monthly reports to the personnel department. On their return they discuss their experience with the personnel manager and make suggestions for improvements to the scholarship programme. For example, Kirishima suggested that the US $1,300 monthly living allowance be increased in order to cover adequately expenses associated with socializing with other Japanese businessmen. He also suggested that the corporation should make a donation to those institutions accepting their employees. He recommended that the company establish a system where candidates' academic levels could be matched with the quality of recipient institutions' programmes. He also advocated more intense English language training at the company's expense before candidates go abroad. Suggestions like these provided by returning employees are examined and, where possible, implemented.

Yasuo Ezawa, assistant director of SECOM's recruitment section and in charge of opening SECOM's Thai office, advocates the expansion of the overseas studies programme to include extended stays in those countries, in addition to the USA, where SECOM has branches, in order to provide employees the opportunity for language and cultural studies. He recommends a closer integration of overseas studies with the company's plan for overseas expansion.

As part of SECOM's short-term employee education programme Iida implemented overseas study tours in 1987. The first was in conjunction with the 'Futa' campaign. The latter was aimed at revitalizing the company and was launched to mark its twenty-fifth anniversary. One hundred and twenty-six employees who excelled in the campaign were selected for a one-week visit to the United States. In 1989 SECOM instigated a business procedure campaign entitled 'BEST Strategy'. From the participants 270 were selected for a two-week visit to the United States to see SECOM's residential and medical security operations.

Iida not only wants to help his young employees broaden their international understanding, in addition to this he has enrolled SECOM in a cooperative programme to assist foreign university students living in Japan. Instigated in 1987 by Keizai Doyukai, or the Committee for Economic Development, a nationwide organization of some 9,000 of Japan's top executives, the programme makes room and board in company dormitories available to these students for a nominal fee to help offset the extremely high cost of living and affords them daily contact with young Japanese workers. At the same time company employees are given the opportunity to broaden their horizons through daily contact with foreigners on a one-to-one basis. Presently about eighty corporations participate in the programme.

SECOM has been involved since the programme's inception when Iida's participation was solicited by Takashi Ishihara, chairman of Nissan Motor and Keizai Doyukai's director. The location of each of SECOM's forty dormitories and the personality of the dormitory director were taken into consideration in the decision concerning placement of the students. At first it was thought that the students would be placed in pairs in a number of dormitories but then it was decided that they might prefer to be together in one group. Seiwa dormitory in Kawasaki city was selected and presently nine foreign students from the People's Republic of China and from the Republic of China (Taiwan) are in residence. Thus, not only does the pro-gramme provide the opportunity for the growth of understanding between Japanese and non-Japanese, it also brings together students from countries where presently no political relations exist. At Seiwa dormitory students participate in a variety of activities including year end and New Year's parties, and sports events.

CHIEF EXECUTIVE OFFICER

With the expansion of corporate activities, Iida felt the time had come in 1976 to realign the top level organizational structure. He appointed Juichi Toda Vice-chairman and, relinquishing his position as president, Iida became Chairman and Chief Executive Officer. Iida was well aware that he had to cross the boundaries that divided the organization into functional areas such as research and development, marketing, sales, finance, and so on and assume responsibility for all dimensions of the business.

> I wanted to broaden the administrative base of the company and thus lessen the negative aspects of the self-centred behaviour often associated with founder/presidents like myself. As president, I couldn't distance myself sufficiently from the corporation. It was difficult for me to give my employees the free hand they needed to grow and to develop their potential. When I attempted to give them the freedom they required I would worry about their progress and would find myself intervening. I would continue to pull back and then be drawn in again. Finally I concluded that I had to let my employees take full responsibility for their work and carry the ball themselves. At the same time, I wanted to be able to move freely and to leave the daily routine to my successor. I wanted to devote my energies to long-range planning in preparation for the corporation's evolution as a vital part of a society in which information communication would be paramount.

Iida's intention was to gain more time to think about network management. He knew that the network business he was planning would face stiff competition. He needed to formulate the scope and direction of the company for future development.

At forty-two, Iida was the youngest CEO of any listed corporation in Japan. It is unusual in Japanese corporations to have a young, active CEO. A study published in the mid-sixties determined that in 1962 nearly 80 per cent of CEOs of major Japanese corporations were over 60 years of age and 50 per cent were over 70.[3]

In assuming his new role, Iida was not prepared to relinquish his authority to his successor. It was understood that, in any position within the company, the authority entrusted to the incumbent could be revoked at any time, if Iida found the managerial behaviour to be inappropriate.

Within this new organizational system Iida appointed Ichiro Tarui president. Tarui had recently come to SECOM from the Nippon Bank. Iida realized that it would be advantageous to the company to bring in a competent executive with expertise and connections in the banking world. A number of individuals with well-known banks had been suggested to him but the choice of any one of them would, Iida felt, link the company too closely with that particular financial institution. He wished to maintain SECOM's neutrality and so looked to Nippon Bank, the central bank, to supply the appropriate individual. Iida felt that bringing in someone from Nippon Bank would elevate the corporation in the public's eye. He consulted with Torao Igarashi, Chairman of the Board of Nippon Bank, who in turn talked with Yasushi Mieno, deputy personnel manager. Mieno approached the 50-year-old Ichiro Tarui, a friend from university days and then a property and facilities manager with the bank. Mieno knew that Tarui was a serious individual whose personality was well suited to the banking world. Iida's influence would 'soften' him, Mieno felt, and help him adapt to the private sector.

Tarui agreed to consider leaving the bank to join Nihon Keibi Hosho. His decision, he said, would rest upon two factors: the results of a thorough medical examination and his feelings when he met Iida face to face. If his health were questionable he felt it would not be fair to Nihon Keibi Hosho to take the job. The medical examination showed him to be in excellent condition so Igarashi arranged for a meeting between Tarui and Iida at which he and Mieno would also be present. Says Tarui:

> I was impressed by Iida's foresight, his broadmindedness, his quick thinking, and his decisiveness. Within fifteen minutes of meeting him I knew I wanted to work with him. His personality reminded me of my good friend Mieno.

Although he changed his title Iida relinquished none of his authority. It was understood among the executives that he would be in complete control of the emerging SECOM group. In executive meetings he presented his ideas and solicited those of others. But, as had always been the case, the final decision rested with him. He would continue to take all the risks and assume all the responsibility. The role of Tarui, as president, and the other executives, was to work towards the accomplishment of clearly defined tasks which were an integral part of the dynamic process towards the actualization of the corporate mission led by Iida's expanding entrepreneurial vision.

8   Innovative ventures

KNOWLEDGE EXCHANGE

The year Iida became CEO he tackled a number of new projects. Continued business success, he believed, partially depended on the extent of the general public's understanding of safety, regardless of whether or not Nihon Keibi Hosho would cater to an expanding popular market in its product/service line. In order to increase public awareness about the importance of security, Iida established the subsidiary company, Security World Corporation.

Security World is responsible for the publication of the bi-monthly magazine, *Security*, under the editorship of Zenji Katagata, a graduate of the Carnegie Mellon Institute and Columbia University and presently chairman of the System Research Centre in Tokyo. The magazine covers such areas as security against theft and other criminal activities and fire, facilities security incorporated into architectural design, and environmental protection. Academic papers, suggestions, transcripts of discussions, explanations of security-related legal information, reviews and bibliographies of security-related books, ethno-historical examples of security in various nations, and serialized security-related science fiction novels are included in the one hundred page publication.[1] The magazine is addressed to public and private policy-makers, researchers and academics.

Iida wished to strengthen further Nihon Keibi Hosho's links with these groups, especially through widening the forum for knowledge exchange in science and technology. He believed that postwar economic development and the nation's future progress depended on the country's ability to advance in science and technology. By the late seventies Japan was facing problems of an aging population, scant natural resources, an unstable energy supply, and environmental concerns. Iida felt that in these circumstances, it was essential, in order to maintain the country's advantageous position in international competition, that advancements in science and technology occur. Through the establishment of the SECOM Science and Technology Foundation in 1979, Iida promoted research and development in security and disaster prevention, and information exchange at both the national and international level in order to ensure the safety of the populous against both human-inflicted and natural disasters. The Foundation began with an initial grant fund of 1.8 billion yen. To date the Foundation has sponsored fifty research projects involving 200 researchers. These have investigated many aspects of security including factory safety, rescue operations, railway accidents, explosion prevention, product safety, health management, as well as many areas directly related to SECOM's business. The Foundation, often in co-sponsorship with academic institutions, research institutes, or newspaper companies, offers symposiums and lectures on fire prevention and earthquake survival in urban areas, the security and information systems within the urban infrastructure, causes of automobile and airline accidents, and computer crime. Presenters are academics, researchers, consultants, and politicians. In order to encourage students to participate in security-based research, the Foundation solicits annually undergraduate and graduate papers on issues related to safety in everyday life written from the viewpoint of the author's field of specialization. Six are chosen for monetary awards. The work of the Foundation is contributing to the development of a holistic view of security and safety.

HIGH TECHNOLOGY AT HOME BASE

A major turning point for Nihon Keibi Hosho in terms of system architecture came in March 1975 with the introduction of an integrated computer security system. With the original SP Alarm System, the on-site terminal at the secured premises was directly linked to the control centre by a private communication circuit. At the control centre, each client was represented by a unit light which would respond when on-site sensors picked up any unusual activity. Monitoring the thousands of unit lights in the control centre twenty-four hours a day was a tedious job requiring unfaltering attention to ensure that no mistakes were made. The SP Alarm had undergone many improvements since its introduction in 1967, but there had been no fundamental change in the monitoring process at Nihon Keibi Hosho's Harumi control centre and its eighteen counterparts throughout the country.

Introduction of the computer security system revolutionized the control centres. Now information transmitted from the secured site no longer registered as a flashing light; rather it passed through a node processor and the details were displayed instantaneously on the computer monitor. Site location was given along with the nature of the problem, for example, whether it was a fire or break-in. With this new system each of the control centre personnel no longer had to watch hundreds of unit lights. Instead each had only to pay attention to his assigned computer monitor. The job of the control centre personnel was thus streamlined and overall response time reduced.

The monitoring capacity of each control centre increased substantially with the introduction of the computer security system and the number of centres remained the same until 1988. In that year twenty-nine additional centres were established bringing the total number to forty-seven and providing one in each of the county's prefectures. The expansion was necessary in order to provide SECOM's rapidly growing clientele with a more regionally focused, multifaceted service made possible through the multiple use of communication channels.

As the number of leased private circuits of Nippon Denshin Denwa Kosha increased with the growing popularity of the SP Alarm System, Nihon Keibi Hosho's leasing expenses rose dramatically even though a portion of the cost was absorbed by the Alarm System users. Iida was unconcerned. He had not yet worked out the details in his mind but he could sense that Japan was at the brink of a rapid shift towards an information technology society. He knew that the larger the number of SP Alarm Systems installed, the greater would be his use of, and therefore control over, communication circuits. He would be able to reach consumers nationwide, offering the same quality services to all at reduced cost. Through expansion of the SP Alarm System he would obtain more circuits than any other company, essential to his goals for business success in the forthcoming information society. A critical step in the realization of these goals was the establishment of a secure base in the country's information communication network.

The leased private circuits could be used in any way the corporation wished. The actual usage of each circuit averaged less than thirty seconds a day. During the remaining twenty-three hours, fifty-nine minutes, and thirty seconds, the circuit was unused. As well, the SP Alarm System was designed so that information picked up by on-site sensors at a location secured by an SP Alarm System could be prioritized and transmitted immediately to the control centre, overriding any other use of the circuit. The rest of the time Nihon Keibi Hosho could use the circuit for whatever purpose it wished. Iida thought it was too soon to share his future direction with his employees. He told them only that he would take all responsibility for corporate risk and instructed them to secure as many contracts for the SP Alarm System as they could. He was fervent in his efforts to market the SP Alarm System. Potential customers abounded. Looking at the major categories of institutional customers alone, nationwide in 1975 there were 245,000 restaurants and stores, 120,000 business offices, nearly 98,000 factories and workshops, and nearly 40,000 hotels, hospitals, and schools.[2]

The use of communication networks for the establishment of intracorporate on-line systems was not uncommon at that time but no entrepreneur had yet conceptualized the idea of a business based on a communications network. Government monopoly of the nation's communication networks effectively deterred entrepreneurial exploration of the possibility of launching a network business. Iida questioned the situation and searched for opportunities. What followed was a series of innovative and, in some cases, inventive ventures.

EXPANDING FRONTIER

At about the time the term SECOM society first appeared in the newspaper, the initial step towards its realization had been taken. In conjunction with the Huntington National Bank of the United States, Nihon Keibi Hosho began research into an unstaffed bank security system, later to be known as the Hanks System. The idea was initiated by Iida on learning that the Huntington Bank was offering twenty-four hour automated-teller service to its customers. It would not be long, Iida realized, before Japanese banks would be offering the same type of service.

On-line systems for financial institutions emerged in Japan around 1965, followed about five years later by a dramatic increase in cash dispensing machines. This was part of a corporate rationalization process necessitated by labour shortages, rising labour costs, and consumer demand for extended service. Initially located inside banks, the cash dispensing machines were soon moved to external locations accessible to customers round the clock. In order to gauge, and at the same time raise interest in security issues, Iida organized a symposium related to the security of financial institutions which were headed in the direction of automated banking services. The response was overwhelming, reflecting the immediate need for integrating bank service with security. The only system available in Japan for this emerging need was that offered by Nihon Keibi Hosho.

Iida first looked into the security of automatic cash dispensing machines and marketed a product called the CD (cash dispenser) Security Pack in 1974. This was followed in the autumn of 1976 by the expansion of the company into an armoured car service for the transportation of cash, stocks, and bonds. Improvements in the CD Security Pack led, in 1980, to the introduction of an expanded system in which control of all aspects of the CD/ATM (automatic teller machine) corner were centralized. By 1983 the entire facility could be secured and functions such as heating, lighting, air conditioning, background music, and entry control automatically operated. These functions could be programmed to operate differently according to day of the week, season of the year, and so forth. The system was adaptable to the needs of organizations varying in size and could be operated on an ad hoc basis as well. Operating in conjunction with theft and fire security systems this system improved greatly customer services.

As well as bank security, SECOM is concerned with the security of cash and valuables in other institutions. In 1983 the company developed the Pythagoras safe primarily for store and office use. It was the outcome of Iida's wish to advance beyond the traditional concept of a safe as heavy and fixed and to design one that could be assembled on the client's premises, yet that would still afford optimum security. Sectional delivery and on-site assembly makes the Pythagoras safe suitable for locations where entry-way space is restricted, a common problem in Japan especially in older stores and offices.

When delivered the safe is in six sections (four walls, a top and a bottom). Each section is composed of six fused, laminated panels resistant to penetration. Once assembled, the safe is equipped with three locking systems and two sensors, one to detect heat and the other to detect vibration. Linked to the control centre, the safe is virtually unpenetrable.

In February 1989 the Ministry of Finance notified financial institutions throughout the country, all of which were moving in the direction of Saturday closure, that security corporations involved in the transport of cash were now authorized to fill cash dispensing machines on Saturdays should the need arise. Until this time this activity had been classified as a teller's job, to be entrusted to no one other than specified bank employees. With the new ruling security corporations were allowed entry into the 'inner sanctuary' of the banking world and the link between them strengthened.

With Nihon Keibi Hosho's electronic security system well

established in 'traditional' fields the company was vigilant for new applications. One such new use for the system was at fish cultivation plants. One of the earliest installations, in the late seventies, was at a trout farm which had suffered a fifty million yen loss when leaves had blocked the excess water outlet causing the water level to rise rapidly and allowing the rainbow trout to swim into the river.

Another early user was an eel farm. A power failure caused by an electrical storm during the night shut down the oxygen blower. By the time the problem was discovered in the morning two hundred million yen worth of eel had died. The eel were not, and could not be covered by insurance. The owner contemplated the posting of round-the-clock watchmen to prevent the recurrence of such an incident but realized that human error was possible. He decided that the money would be better spent on electronic surveillance and consulted Nihon Keibi Hosho. As with the trout farm, with minor adjustments in the sensors, the system was readily adaptable to the new situation.

FIRE EXTINGUISHING SYSTEM

The prevalence of wood and paper used in construction in Japan, the proximity of buildings to one another, and open flame household gas and oil stoves easily overturned by human carelessness and frequent earth tremors have combined to make Japan a country in which fires are commonplace. The danger is magnified by congested roads which increase the response time of firefighters.

Fire has always been a primary concern for the Japanese. In feudal times those responsible for starting fires, even accidentally, were exiled from rural communities. The devastation wrought by fires in urban centres has been tremendous. During the Edo period (1603–1868), for example, more than one hundred major fires ravaged Tokyo. The worst of these occurred in 1657, 1772, and 1806 claiming 107,000, 14,700, and 50,000 lives respectively. By comparison, London's Great Fire of 1666 which burned for five days and razed most of the city had no known casualties. Ironically, during the Edo period fires were called Edo no hana (flowers of Edo). The Shogunate attempted to improve Edo's fire defences through the organization of fire brigades and city planning. The Meiji government incorporated fire prevention and extinguishing activities under the police branch of the Ministry of Home Affairs in 1880. The Tokyo fire department was equipped with up-to-date foreign-made equipment including British hand pumps and American pumper trucks. At the outbreak of World War II thirty-six of Japan's urban centres had public fire stations employing some

30,000 people.[3] However, the aforementioned urban conditions meant that the firefighters were almost powerless against the air raid fires towards the end of the war.

During the Allied Occupation, GHQ (General Head Quarters) separated the fire and police departments and created the Fire Defence Agency within the Ministry of Home Affairs. Among the many factors that prompted GHQ to take action was the fact that the beer halls and cabarets opened at their request for the exclusive patronage of Allied soldiers were lacking adequate fire protection and were in constant danger of burning down.[4] The formation of the Fire Defence Agency was a step in the right direction but, despite the Agency's best efforts, fire remained a major hazard. Today 70 per cent of Tokyo fires are associated with buildings and 60 per cent of these with residential structures.[5]

Soon after establishment, Nihon Keibi Hosho began to participate in fire prevention activities as part of its guard duties and, a little later, in the development of a fire alarm system. System subscribers often expanded from individual users to entire communities. In 1969 for example, Unatsuki, one of Japan's 2,000 hot spring towns, decided to upgrade its security measures in order to attract more visitors. Town officials, hotel owners, and police and fire administrators met with Nihon Keibi Hosho's regional manager to discuss a total security package. Initially twenty-three hotels in the area were provided with round-the-clock surveillance. Soon after a number of homes were included in the security package. Eventually most of the town was protected by a human-based system.

Before long Iida saw the need for a highly reliable and effective automatic fire extinguishing system. The company had performed well in the area of theft prevention, developing and marketing a reliable detection and alarm system which propelled Nihon Keibi Hosho to the number one position in this field.

> At the same time, despite the fact that we have a social responsibility towards all areas of security, our efforts had been concentrated on theft, at the expense of fire prevention. Not turning our attention to the development of an appropriate fire prevention system would lessen the possibility of the future development of the corporation. To me, it would also represent a corporate betrayal of the confidence and expectations society had towards us as a leading company expected to provide total security.

Fire security has three dimensions: fire prevention, fire containment and extinguishing, and fire escape. The general trend at the time was after-fire feedback control. We took a more positive approach seeking ways to prevent fires from starting and, if this failed to extinguish them immediately. From this starting point we developed an innovative fire detection and extinguishing system.

The system consisted of heat and smoke sensors, nozzles linked to halon gas or dry chemical tanks, and a computer controller which would sound the alarm and cause the gas or chemicals to be released.

Traditional fire alarm and extinguishing systems relied on a manually activated alarm to alert persons in the area of the fire, their subsequent notification of the fire department, and the arrival of fire fighters to extinguish the fire. Early detection, successful communication, and rapid arrival of fire fighting equipment are essential if the fire is to be controlled. Automatic water sprinklers, often a part of this traditional system, cannot be readily checked for dependability. Also, when the sprinklers are activated, the damage caused by the water can be considerable.

The system Nihon Keibi Hosho developed did not depend on water sprinklers. Halon gas 1301 was the universally effective, and initially believed to be nearly harmless, extinguishing substance utilized in Nihon Keibi Hosho's system. Halon gas was less likely than water or foam to conduct electricity. Odourless and colourless, it did not hamper visibility and did not create obvious damage. Because of its expense Halon gas, until 1974, had only limited application. Its widespread use was considered prohibitively costly. Nihon Keibi Hosho developed a system which permitted inexpensive integration of the gas. It found wide application in computer centres, communication centres, power plants, libraries, museums, enclosed parking lots, and airplanes. When this system was installed the customer had to pay only a predetermined monthly fee whether or not the system was activated. It would be sometime before the environmentally destructive properties of Halon gas would become a public concern.

Under the surveillance of the on-line computer controlled system, when fire broke out the system would be activated immediately and beat engineers automatically called to the scene. The system was checked periodically and any part that malfunctioned replaced with its most advanced counterpart. Once the system was developed it had to be named. In-house discussion resulted in the choice 'Fire Alarm Full Automatic'. Because of the repetition of the initials FA this was abbreviated to FA2. The name that was adopted was SECOM FA2.

Prior to the commercial launching of the system Iida called a meeting of the company executives, branch managers, and sales, installation, and maintenance managers to describe the significance and function of this system and the importance of its thorough after-care. The first system was put on the market in July 1974. This was the prototype for further development. The following year the company introduced a more compact version.

Household, commercial, and institutional kitchens were a major source of fire in Japan. In the mid-1970s the Fire Defence Agency of the Ministry of Home Affairs presented standards for automatic fire extinguishing systems for kitchen safety. Major cities, including Tokyo, Osaka, Nagoya, Kyoto, Sapporo, Yokohama, Kobe, Okayama, Hiroshima, and Fukuoka, introduced a series of by-laws governing the compulsory instalment of automatic fire extinguishing systems. Nihon Keibi Hosho, as well as a number of other corporations involved in their manufacture, worked together with the various government departments to ensure that the regulations were realistic. By 1977 Nihon Keibi Hosho had developed a flame sensor and two years later it introduced a system to extinguish kitchen chimney fires. The new series carried the name Tomahawk, suggestive of its ability to function as a chemical axe and halt the fire's advance.

With these systems on the market Nihon Keibi Hosho was in a position to offer significantly improved hotel fire protection. Hotel fires are a continuing problem in Japan. Recent disasters include the Kawaji Hot Spring Prince Hotel fire of 1980 which claimed forty-five lives, the Hotel New Japan fire of 1982 in Tokyo which killed thirty-two, and the 1986 Atagawa Hot Spring Hotel Daito-kan fire in which twenty-four died. These hotels, like most modern Japanese hotels where fires have occurred, had fire alarm systems. The alarm systems, however, may have malfunctioned and the alarm did not go off or, because of recurrent false alarms and customer complaints about the disturbance, management may have shut off the alarm system. Another possibility is that the alarm sounded but there was no appropriate on-line emergency communication network linking the hotel premises with the fire station and thus response time was too long.

Iida realized that Nihon Keibi Hosho's work on the development of effective fire alarm and extinguishing equipment could be advanced significantly through consolidation of the two technologies. In 1985 the company became the largest shareholder (33 per cent) of Nohmi Bosai Limited, Japan's leading fire extinguisher manufacturer, and later purchased stock in Hochiki, an alarm company, thereby facilitating a committed technical exchange. Early research results were developed into an integrated fire prevention system with the capacity for smoke accumulation and heat retention analysis and the system was marketed in 1986. In this system, strategically placed sophisticated sensors and automatic extinguishers are linked to the hotel's control terminal which is in turn connected on-line to the control centre and thence to the appropriate fire and police stations. The corporation claims that this system has reduced the number of false alarms to seven one-hundredths of what they were previously.

Recognizing the difficulty in operating a traditional fire extinguisher, especially for the elderly and for many women, Iida saw a business opportunity in the need to redesign the extinguisher. The outcome was a stationary canister with a ten metre flexible hose the nozzle of which can be operated with one hand. Removal of the nozzle from the canister alerts the control centre to the outbreak of fire, possibly even before the sensors do so, and the fire defence system is activated if the situation requires it. Introduced into the market in 1989, as the Tomahawk Mach, the extinguisher is easily activated even by children and is appropriate for extinguishing all types of fire. Incorporating maximum effectiveness with convenience, it is widely used in residential structures, offices, and factories. The development of the Tomahawk Mach is an example of business development based on product modification.[6]

ENTRY CONTROL

Entrance and exit security control was another area which attracted Iida's attention. Traditionally, locks and keys were relatively unusual in Japan. The earliest mention of keys in Japanese recorded history is in the list of materials used in the construction of Horyu-ji, Japan's oldest extant temple, founded in Nara in 607. A key was prepared for the gate and also for the repository for statuary and ritual equipment. The idea of a lock and key and most likely the fabricated locks and keys themselves came from China. After its introduction, there was virtually no development in 'key technology' until the Edo period (1603–1868) when limited use was made of keys in the storage houses of wealthy merchants. These were the first Japanese-made keys. They were artistic in design, often in the shape of a drum or a shrimp, but of little practical value. Storage houses were frequently broken into with little effort. A symbol of wealth, the key functioned similarly to the seal, signifying that the door was not to be opened but affording little actual deterrence. Therefore, it was functionally different from its western counterpart.[7]

Even towards the end of the Edo period few commoners were familiar with keys. In contemporary rural Japanese society there are still many homes which are without keys, in part because of the close-knit communities where neighbours look out for each other. The widespread use of keys by city dwellers has come about only since the end of World War II. Until then the extended family lifestyle ensured that at least one person would always be at home to safeguard the premises and open the door, which could be locked from the inside only, for family members. On the rare occasions that all family members would be out at the same time a neighbour or relative would be called upon to stay in the house.

Moreover, owners of traditional Japanese houses saw little point in adopting keys because the door was only one of numerous ways through which the house could be entered. Access through easily removed sliding doors, via the crawl space under the house and up through the tatami floor, or even by the removal of roof tiles were alternatives available to any burglar.

Societal changes over the last forty years, however, have had a major impact on key usage. With the rise of the nuclear family the likelihood of having someone always at home has diminished.[8] New construction methods and materials have resulted in more secure dwellings with fewer easy entry points. As well, where easy entry also meant easy escape, if discovered, burglars in newer homes are likely to be trapped in an enclosed room with no escape route open to them, creating a dangerous situation in case of confrontation. The considerable increase in affluence among average individuals as a result of Japan's rapid economic growth has vastly increased material wealth making almost every home the worthwhile focus of a burglar's attention. For all of these reasons more sophisticated locks and keys have become necessary especially for urban dwellers.

Architectural and lifestyle changes as well as the growing importance of privacy and protection in the newly emerged information society necessitated the development of sophisticated entry control systems bearing little resemblance to their predecessors. In the institutional context, where keys have been important for security for several decades, until recently technological development had been largely ignored. As Japan developed as a technological leader the protection of materials and information became more important and a market emerged for reliable entry control systems. Computer centres, laboratories, communication centres, control towers, executive offices, as well as traditional sites such as mints, jails, airports, banks, and professional offices, to name a few, all required a high level of entry control.

In 1976 Iida introduced the security lock system prototype having as its basic components a magnetic card reader, push button code reader, and timer linked to an electronic controller, intercom, camera monitor, and a printer which recorded entry and exit activity. The system was improved gradually. In 1987 the company marketed an 'optical' entry system which uses infra-red rays and the following year introduced a biometric security system capable of recognizing individual fingerprints. Integrated with personal computers, this system is more dependable and economic than its forerunners.

Privacy intrusion can occur by telephone as well as by physical entry. SECOM's answer to the problem of unwanted telephone calls was the development of a telephone with a built-in 'echo' function. At the push of a button, the caller's voice is recorded for a few moments then automatically played back to him/her. Knowing that no one was listening and that the voice has been recorded will often serve as deterrent to future calls.

NUCLEAR SECURITY

Japan's energy and natural resources are insufficient to meet its industrial requirements. With increasing dependency on imported oil from 1960 on, Japan's self-sufficiency in energy has declined. The oil crises of 1973 and 1978 temporarily paralysed the nation and highlighted the importance of developing alternative energy sources. Long-range energy research and development programmes have become a high priority.[9]

Among several alternatives, nuclear energy is perceived to hold the greatest potential. Nuclear energy was endorsed especially strongly in the late 1950s and throughout the 1960s. In 1955 a law was enacted stipulating that Japan's development and use of nuclear energy is to be for peaceful purposes only and is to be conducted in a democratic, independent, and safe manner with all results publicized. In the 1980s in Japan, as in the United States, the perception of coal as a cost-effective fuel for generating electricity increased.[10] Nevertheless, Japan continued to develop its nuclear power generation capacity and today Japan ranks third after the United States and France with thirty-five commercial nuclear reactors. As of 1988, 15 per cent of Japan's electric power was nuclear generated. Paralleling the rise in nuclear power generation has been the development of related industries such as uranium mining; refining, enrichment, processing, and reprocessing of nuclear fuels; and nuclear waste disposal.

Entry into the nuclear age was far from smooth, however. The only people to have experienced nuclear attack, Japanese opposition to nuclear power in any form has dominated the ongoing debate on nuclear development and has been dubbed the nation's 'nuclear allergy'. The sixties were marked by anti-nuclear demonstrations as the country cautiously moved into the commercial production of nuclear generated electricity.[11]

Against this background the need to ensure the absolute safety of nuclear energy was apparent. Two government bodies, the Nuclear Energy Council and the Nuclear Energy Safety Council, had been established to supervise the development and utilization of nuclear energy and to maintain safety. Clear lines of authority, well trained and highly skilled personnel, reliable communication systems, and specialized equipment would have to be in place. Government and private electric power companies with an increasing reliance on nuclear energy had been exploring this issue since the first oil crisis hit in 1973.

With the technology and human-based support structure Nihon Keibi Hosho had developed, Iida had the confidence to involve the company in the design of large-scale security systems. In July 1977 Nihon Keibi Hosho participated in the establishment of JNSS or Japan Nuclear Security System, a joint venture with Japan's three leading private electric power companies: Tokyo Electric Power, Kansai Electric Power, and Chubu Electric Power. The function of the corporation is to carry out research and development into the design, installation, and maintenance of nuclear plant security systems and the safe transportation of nuclear fuels. Here all of SECOM's technological know-how acquired to date has been used and is expanding constantly as the systems undergo modification and improvement.

JNSS is responsible for the security of ten of Japan's fifteen nuclear plants which house twenty-nine of the country's thirty-five commercial use nuclear reactors. All ten plants are located on Honshu, Japan's 'mainland'. JNSS is also responsible for the security of nuclear fuel processing plants including the country's two largest – Mitsubishi Nuclear Fuel and Japan Nuclear Fuel. In addition, JNSS oversees security at Mitsubishi Heavy Industry Nuclear Research Plant, thermal electric power stations, and LNG plants. To date, no security incidents have occurred within JNSS's jurisdiction.

HOUSEHOLD SECURITY

Many variations of the SP Alarm System have been developed since it first came on the market. Until recently its primary use was in institutions. Its most common application is in office buildings, factories, retail stores, and schools. From the outset Iida's intention was to provide security economically for as wide a spectrum of users as possible. He felt that while the company had made significant advances into the field of institutional security, much needed to be done in the area of home security.

To be sure, Japan's low crime rate is still the envy of the world. Statistics for 1984 indicate 1.47 murders per 100,000 population compared with 7.2 in the United States, 3.24 in Britain, and 4.51 in West Germany. Burglaries and petty thefts are relatively common but robberies involving assault are much lower than in the West: 1.82 cases per 100,000 population in Japan compared with 205.38 cases in the United States, 50.02 cases in Britain, and 45.77 cases in West Germany.[12] But societal perception was changing. With the basic physiological needs of the population at large being met through postwar prosperity, there was increasing interest in social amenities including the attainment of peace of mind through physical security. Increasing numbers of business and holiday trips, the upsurge in cottage or second home ownership, and the widespread entry of women into the labour force resulted in an increasing number of dwellings left empty for an extended period of time. As well, Japanese returning from European and American postings brought with them an increased awareness of the desirability of residential security measures.

Despite the apparently effective police force and the relatively low crime rate, a 1980 survey based on the responses of 6,002 residents of selected districts of sixteen urban centres in Japan indicated that all felt that they might someday be the victim of a crime. Sixty per cent feared robbery and 60 per cent of female respondents, regardless of age, feared molestation.[13] In addition, the Japanese also feel insecure about traffic accidents, fire, and earthquakes and other environmental disasters. In response to changing needs, Nihon Keibi Hosho developed Japan's first on-line home security system. It was introduced to the market in January 1981 under the name 'My Alarm' and by 1989 contracts numbered 25,000.

In this system sensors strategically placed throughout the house will check continuously for fire, gas leaks, and illegal entry. If an irregularity is detected the computerized controller will sound the alarm and simultaneously transmit the information to the control centre which in turn relays it to the depot nearest the premises. From here the appropriate beat engineer is immediately dispatched to the site and the proper authorities notified. The combination of computer controlled on-line real time response and the human element in the system makes this service distinctive.

'My Alarm' permits adjustment according to the situation, for example night-time, when the house is occupied, and when it is unoccupied. The home is divided into a number of self-contained security areas which can be activated or shut off according to the user's requirements.

An auxiliary system for 'My Alarm' is 'My Doctor' introduced in November 1982. The latter is an emergency alert buzzer system to be activated manually by the wearer when a problem arises. As the Japanese population ages and the nuclear family replaces the extended family adequate care of the aged has become a widespread concern. Iida was acutely aware of this problem on a personal level as his mother approached 90 years of age and he realized that SECOM's technology could be harnessed to contribute to its alleviation.

The forty-gramme alarm button SECOM developed (which also comes in a waterproof model) is worn as a pendant around the neck and is activated by squeezing. Wearer activation of 'My Doctor' registers in the control centre. A beat engineer is dispatched to the site immediately while the control centre attempts to contact the wearer by phone to determine the exact nature of the problem. Of course if the problem is a fire, break-in, or gas leak this information will already be registered through 'My Alarm' and the appropriate emergency services notified. If there is no response to the telephone call, a district welfare commissioner is sent to the site and neighbourhood friends are alerted.

In addition to the widespread adoption of 'My Doctor' on the part of individual households two cities, Kawasaki and Wakayama, initiated pilot projects to determine the usefulness of installing the system in the homes of their elderly bedridden citizens who live alone. Other municipalities are following their example in response to the growing nationwide phenomenon of the bedridden elderly. A Health Ministry survey has found that the proportion of aged people in Japan who are completely bedridden is six times higher than in Denmark and three times higher than in Britain.

A 1988 Wakayama city survey revealed that among its 402,000 inhabitants 5,700 over the age of 65 lived alone. Of these 1,000 were ill, half of them bedridden. The city viewed the provision of 'My Doctor' as an initial step in fulfilling the minimum responsibility of a safe environment for the elderly. In Wakayama help can be on the scene within twelve minutes, dispatched from one of four depots strategically located throughout the city. In 1988 the system was credited with saving six lives and its expansion deemed worthwhile by the city council. In that year Wakayama city budgeted for the installation of the system in 140 homes with an additional 140 to be installed in 1989. Presently the system is installed in 400 city homes. The intention is to make the system available eventually to all 500 of the city's bedridden citizens. The expense is increased because a prerequisite for 'My Doctor' is the installation of 'My Alarm' to provide the necessary infrastructure for its operation.

Mutsu city in northern Honshu installed 'My Alarm' and 'My Doctor' in the homes of thirteen of its elderly citizens in the autumn of 1989 with the intention of eventually making the alarm system available to all city residents over the age of 65 who live alone. One important contribution of 'My Doctor' is the peace of mind it brings to users. Seventy-five year old Yone Ogasawara started to use 'My Doctor' in October 1989. She was quoted as saying:

> I suffer from asthma and heart problems and have no family members close by on whom I can call should I have an attack. I used to worry a lot about what might happen to me, especially at night. I was dependent upon the good will of my neighbours but with 'My Doctor' I feel a great sense of relief and vitality.[14]

By 1990 more than ten cities throughout Japan had integrated 'My Doctor' into their programme of care for the elderly. At present the system is installed in Japanese homes in ninety-two cities, towns, and villages.

Iida's mother was one of the first users of 'My Doctor'. At her advanced age Iida was concerned about her ability to understand how to use the device. When he queried her she replied, 'Yes, I understand. If I get into trouble I push this button and someone will come and help me.' This was exactly the response for which Iida had hoped. It was important for 'My Doctor' to be perceived as user-friendly and simple for anyone to operate. The ease of operation of any piece of high-tech equipment is, Iida feels, the key to capturing a large market share.

In addition to 'My Doctor', other options that may be purchased with the home security system include a camera integrated intercom system with TV monitor, electronic door locks, and telephone activation of the security system, lights, heat, air conditioning, and other household equipment. With regard to institutional security, Iida saw in the holistic integration of various aspects of building management the opportunity for business expansion.

TOTAL BUILDING SERVICE

With the increasing number of large office and factory complexes the necessity of securing all facets of the building environment has become apparent. Traditionally, building owners would contract out various tasks including security against theft and fire, and air conditioning, heating, electrical, plumbing, and elevator maintenance and repair. Most of these jobs are labour intensive.

Iida saw an opportunity for Nihon Keibi Hosho to offer a package service that would oversee facilities management in its entirety. Just as Iida moved from a labour intensive security guard operation to a computer integrated service in the sixties, in the late seventies, he applied a similar idea to building management. Iida argues that only corporations with the ability to integrate communication networks with manpower services could offer effectively this type of service. SECOM's building facilities management package became available in March 1978.

Two years later, in June 1980, Iida launched SECOM Systems Company, Nihon Keibi Hosho's engineering arm. It was intended to meet widely divergent, large-scale security needs, including building management. The following month he established SECOM Maintenance to provide services related to building design, equipment and facility systems sales, and maintenance. Iida had been gaining a competitive advantage based on product differentiation and low costs achieved through scale merit. Now, through the establishment of an engineering and maintenance arm, he was striving to increase Nihon Keibi Hosho's competitive advantage by utilizing the interrelations between affiliated business units and to create enterprise differentiation.

With SECOM Systems Company and SECOM Maintenance providing the service support network, Iida was ready to make building management a viable component of SECOM's services. Among the first clients were office building landlords. With the shortage of land and Japan's high economic activity, office rental became common in the larger cities. In addition to increasing security costs brought about by rising wage rates, building owners were confronted by security problems arising from diverse facility usage. Different occupants, for example corporate branch sales offices, jewellery shops, financial institutions, law firms, and restaurants, required different types of security at different times of the day and night depending on their hours of operation. Securing areas of common usage, such as lobbies, stairs, washrooms, and elevators was of particular concern as were emergencies which might arise with equipment such as boilers, air conditioners, and so on during the night.

In order to meet the demand for a security system that would overcome these problems, SECOM developed a machine-focused system capable of accommodating the needs of individual occupants while at the same time securing the total premises. The system was named the SP Alarm G and became the core from which the concept of total building management expanded. An appropriate sensor was placed in each area of the building and linked to the computer terminal by private circuits. In case of emergency the SECOM employee staffing the building's control centre would notify immediately the nearest depot and, if necessary, the fire and/or police department. The depot would dispatch a beat engineer (or engineers) to the site.

Compared with traditional, guard-based security which results in some areas being unsupervised at any given time and creates the risk of human error, this new system provided more reliable service with area, rather than point, coverage. At the same time costs were lowered with the reduction in recruitment and training of security personnel. The presence of the SP Alarm G within a building was also a positive feature in attracting tenants. As well as in office buildings the system also found application in factories, warehouses, schools, distribution centres, research institutes, condominiums, financial institutions, shopping centres, and underground commercial complexes.

SECOM's innovative view of network architecture conceptualized the network as an entrepreneurial tool, a means to coordinate resources. What made the new system unique was its divergence from the traditional concept of building management which required an in-house computer and the presence of building operations and maintenance personnel. The computer control centre was removed from within the building and integrated into SECOM's main control centre. SECOM's established, remote controlled, centralized communication system – the manifestation of corporate effort and achievement – became the resource upon which the company was able to capitalize in order to develop an innovative new service which provided 'external' building management. The externalization of various activities resulted in a reduction of facility and security administrative overheads.

Iida believes the security company has an important role to play in evaluating risks and advising clients on the best course of action to contain them. SECOM personnel provide comprehensive security surveys which point out not only the obvious risks but also areas which may suffer major losses in the future. The idea upon which total building security is based is rooted in the concept of risk management: viewing the risks as a whole then finding ways of eliminating, reducing, or living with them. Such risk analysis forms the basis for loss prevention and limitation strategy with regard to building security.

In 1979 Nihon Keibi Hosho developed a building energy conservation system in which computer analysis of temperature and humidity data permitted automatic regulation of a building's interior climate. The corporation's intention was to reduce annual energy costs by between 10 and 20 per cent. The fee for the service would amount to only a portion of these savings; the user would retain the remainder. In 1981 this system was incorporated into a comprehensive total building control system. Called SECOM Totax-T, this system falls within the movement towards the design and construction of 'intelligent' buildings. The infrastructure of these buildings facilitates state-of-the-art office automation, automated building facilities management, shared database usage, complete security, and interaction with external information networks.

Clients are sought not only among the owners of existing buildings already using SECOM's security services but also among those about to construct new facilities. In this age of 'intelligent' buildings the hardware for SECOM's total service will be installed during construction. With this system the premises are linked to SECOM's central computer and secured round the clock. The system is supported by what Iida calls 'three muscles': beat engineers, technical engineers, and maintenance engineers. Beat engineers are dispatched to the site by the control centre in the case of fire or illegal entry. Technical engineers are sent if there is a facility malfunction. They will identify the emergency steps to be taken and participate in returning the system to normal. Based on maintenance records compiled by and stored in the control centre maintenance engineers make routine maintenance checks of the facilities.

SECOM argues that its on-line building systems management service can reduce facilities maintenance and security costs by 40 per cent. All of the building management data, including, for example, equipment usage, electricity, water, gas, and oil consumption, and maintenance and repair records, together with the associated costs are reported to clients monthly, while bill payment is handled by the security firm. The intelligent buildings information system incorporates LAN (Local Area Network) and provides holistic integrated building management beneficial to both tenants and building owners/ managers.

9 Vertical integration

PRODUCTION CONSOLIDATION

In the beginning, having very little in the way of physical resources, Iida had no choice but to tap external sources. In addition to domestically produced sensors, foreign-made sensors were also used in the early stages of the development of the SP Alarm System. Ademco sensors, imported from the United States until 1969, were used to detect the opening of doors and windows. Sensors for safes were purchased from the British firm, Securitas, until 1971. With both products delivery was difficult to coordinate and there were problems with post-installation maintenance.

Microelectronics was playing an increasingly dominant role in Nihon Keibi Hosho's security infrastructure where microcomputers and semi-conductors incorporating VLSI (very large-scale integration) were crucial. The seventies saw an exponential leap in the capacity of microprocessors and memory chips accompanied by a rapid fall in price. For example, Japan's first imported microprocessors cost about $1,000 each in 1971; by 1979 Japanese-produced microprocessors were selling for just over a dollar apiece.[1] Recognizing the speed at which the microelectronic field was advancing and being in a position to take advantage of economies of scale, Iida felt ownership of production facilities was essential to maintaining control and bringing about planned corporate expansion.

Until 1977 production of Nihon Keibi Hosho's security-related equipment had been contracted out. About a dozen electronics manufacturers including such major corporations as Matsushita Electric Works, Chino Works, Nippon Battery, Hitachi Electric and Oki Electric produced sensors and controllers according to Nihon Keibi Hosho's specifications. At about this time, writing in *The Economist*, Norman Macrae stated:

It is gradually becoming clear that production is no longer a source of economic or political power, and may indeed now become a source of economic and political powerlessness. It is easy for an organization to take action against sub-contractors by cutting of contracts; it is no longer easy to pass down orders to direct employees.[2]

Contract manufacturing was prevalent in Japan. While contract production capitalizing on existing technology had provided Nihon Keibi Hosho easy entry into the production scene, it presented some problems: production cost control, product quality control, the accumulation of production know-how, technological transfer within the company, the protection of secret product information, and the coordination of research, production, and sales.

To solve these problems Iida had two choices: he could establish his own entirely new production facility or Nihon Keibi Hosho could absorb an existing company capable of meeting its production needs. Iida chose the latter option as it presented the most rapid means of entering the production field.

Among Nihon Keibi Hosho's suppliers was Oki Electric Industry, a leading and long-established manufacturer of communications equipment specializing in telephone communication terminals, switchboards, computers, printers, sonars, and, more recently, semi-conductors. By the mid-seventies Oki Electric had been riding on the coat-tails of the government for some years. It suffered from low morale and lack of innovative management. To rectify the situation the company began a programme of rationalization that included the dismissal of a large number of permanent employees and the restructuring of production needs. This latter strategy included the sale of the Shiroishi factory in the Tohoku district located in the northern region of Honshu.

The Shiroishi factory was a branch plant of Tohoku Oki Electric, an affiliate of Oki Electric, and was producing telephone and switchboard components for Oki Electric. Because Shiroishi had the necessary technology and the capability to accommodate the requisite multi-product small batch production, the factory's excess production capacity was contracted by Oki Electric for the production of Nihon Keibi Hosho's SP Alarm System components. Nihon Keibi Hosho purchased the plant complete with equipment and 270 workers.

The name was changed from Shiroishi Oki Electric to SECOM Industry, the factory renovated and a computer controlled manufacturing system installed. Naoshi Narikawa remained as president of Nihon Keibi Hosho's newest affiliate. Quoted in an in-house publication, shortly after the Shiroishi factory was taken over by SECOM, Narikawa said:

> We had been involved in the manufacture of communication equipment. Through provision of high quality equipment we contributed to the improvement of communications within society. We had established an efficient production system and were ready to embark on the manufacture of security equipment for SECOM.
>
> As Shiroishi Oki Electric we had played a role in the manufacture of a segment of an extensive communication system but as SECOM Industry we are responsible for manufacturing a total security system. Thus we are able to relate ourselves to the total product/service our company is offering. Workers are now involved in a variety of production processes and in this way have less exposure to the monotonous and repetitive tasks which are often associated with subcontract work for large corporations.

CORPORATE NERVE CENTRE

A company whose business is security can hardly overlook its own security. As the number of contracts increased the need for a corporate control centre to support the nationwide on-line security network became apparent. Some years before, at the request of an acquaintance, Iida had purchased land in Hachiman-yama in the Setagaya ward of Tokyo. In financial difficulty, the owner was moving his textile business to the United States (where he subsequently became a successful manufacturer of ladies apparel) and needed to sell the property to offset his debts. Although he had no immediate use for the land, Iida knew that it was well situated both geographically and geologically. This, he determined, was the site on which to build the SD (System Design and Security Development) Centre where all of the company's databases would be stored.

Much of the Tokyo area consists of a thick layer of loose volcanic ash and/or alluvial silt making structural foundations unstable. Earthquakes cause tremendous damage. The most severe earthquake in living memory was the great Kanto Earthquake of 1 September 1923 with a magnitude of 7.9 on the Richter Scale. Two hundred and fifty thousand buildings were destroyed or severely damaged by the shockwaves and nearly as many again by the ensuing fires. One hundred and forty thousand people were dead or missing. Acutely aware of the potential earthquake hazard, Iida knew that the site for the SD Centre had to be geologically stable. Hachiman-yama was ideal. In this area solid bedrock underlies eight metres of volcanic ash.

In November 1978, construction was completed on a five-storey building with its foundation resting solidly on bedrock, as nearly earthquake-proof as the technology of the day would permit. The computer control centre, SECOM's nerve centre, is located in the building's windowless basement. There is a double quake-proof floor. The entire SD Centre is protected by state-of-the-art security equipment. By 1989, the SD Centre housed two IBM 3083 computers and one IBM 3090 computer for security and information processing. These computers are linked to seventeen regional control centres throughout Japan which in turn, via 327 access points (node processors), connect them to approximately 200,000 Computer Security System terminals gathering information from fifteen million sensors located at guarded premises throughout the country. Approximately 20 per cent of these are private homes, the other 80 per cent are industries and private and public institutions.

While the SD Centre monitors activity throughout the nation, it is directly responsible for the Tokyo region controlling the work of fifty-one depots, 120 patrol cars, and about 200 beat engineers. In the daytime each of the five terminals in the computer room is staffed by one controller but at night the personnel at four of the terminals is doubled to handle the increased workload. Only the chief controller works alone. The terminals are monitored around the clock seven days a week.

SECOM

Iida had introduced the term SECOM in 1973 but had met with considerable resistance to his wish that it replace Nihon Keibi Hosho as the corporate name. The name SECOM had been chosen to reflect accurately the company's emerging business activities which made extensive use of information communication computer technology. To Iida, SECOM evoked an image of the direction in which the company should be heading. An innovative, forward-looking corporate self-image, symbolized by the term SECOM, would help optimize corporate performance. With the company's growing involvement in the international arena and the expansion of corporate activities into new information services, Iida decided in 1983 that the time had come officially to change the corporate name to SECOM.

SECOM had been used internally for ten years. The extensive permeation of the name throughout the corporation coupled with the changing nature of security activities had caused the initial resistance to the new name to evolve gradually into employee agitation to make the name change official. Thinking back, Iida recalls:

> This important change was made not primarily because SECOM was easier to say or easier to remember but because the new name reflected the new and expanding business frontier into which the company had moved. We had developed from a security guard business to a security or safety industry and were on our way to becoming a 'social system industry'. This industry would seek the security of society in the expansive sense through the medium of the information network which would include safety/security systems, the company's traditional business focus.
>
> The name change would rid the company of what I call the 'grime' that had built up over the years from a sense of comfort and satisfaction with the status quo. We started with nothing and through strenuous effort the company began to take on a distinct shape. Gradually, as the corporation grew, the corporate atmosphere shifted from achievement to maintenance. Employees were less concerned with challenge than they were with keeping things as they were. This attitude cultivates ineffective employees including a growing number of ineffective executives. This in turn destroys a vibrant corporate culture and stalls corporate progress, leading eventually to the corporation's decline. My intention in the name change was to flush out the 'grime' with which the old corporation was infused.

The decision to eliminate the name Nihon Keibi Hosho and use only SECOM had been made. What remained was how to implement it effectively. For this purpose Iida struck a corporate identity committee comprising representatives from a number of different departments including corporate planning, public relations, advertising, and employee training. The committee's coordinating office was located in the company's general affairs department.

The committee was mandated to clarify the intended meaning of the term SECOM, to develop an externally focused strategy for the visual representation of the term in a variety of areas, and to mount a programme of employee education to promote understanding of the corporate name change.

The committee distilled the corporate philosophy into ten statements:

1 SECOM creates a culture with security at its core.
2 SECOM is innovative, always seeking new challenges.
3 SECOM initiates change; everyone participates in this reform.
4 SECOM promotes concentrated thought and rapid response.
5 SECOM cherishes determination and clear, systematic thinking.
6 SECOM seeks what is right and rejects compromise.
7 SECOM provides safety.
8 SECOM provides peace of mind to clients.
9 SECOM values its professionalism.
10 SECOM challenges future possibilities.

Directly and indirectly, Iida had been stressing these ideas for years. The committee synthesized and abbreviated Iida's theories to ten short statements. Collectively, these statements would become a cornerstone to SECOM's corporate goals, objectives, and strategic plan for capital and human resource allocation. They were termed a morale harbour and supporter. Just as a ship will return to its point (harbour) of departure, so when a SECOM employee reached a dead end as he moved towards the unexplored business frontier he could refer back to these fundamental tenets.

These ten cardinal points continue to play an important role in SECOM's development. Masaoki Kojima, vice-chairman of SECOM's Board of Directors joined SECOM in 1985, coming to the company on his retirement from his position as a senior executive director of Marubeni Corporation. Says Kojima:

> Corporate progress is dependent upon the creation of new corporate culture capable of critical assessment of present corporate activity in the light of future progress. No doubt in the future we will face types of managerial problem never before encountered. Employees' understanding of the reason for the existence of this corporation as initially conceived by Mr Iida is of eternal importance.

The ten statements formulated in 1984 with explanations of their intended meaning could have been communicated to employees through in-house publications. Instead Iida chose to meet personally with as many of the staff as possible to discuss SECOM's cardinal points. Working with groups of sixty at a time, eventually Iida met with 3,600 of SECOM's 9,000 employees who in turn communicated the message to others. Because of the effort invested in orientation and education, when the name change was announced on 1 December 1983 the transition was smooth and immediate. Internally, a one-line memo was circulated to all departments stating that Nihon Keibi Hosho henceforth would be known as SECOM.

In addition to the name of the parent corporation, the names of nine associated and two affiliated corporations as well as the names of related organizations such as the union had to be changed. One of these changes, however, was in reverse. The name of the company's printing centre, the affiliated corporation SECOM Printing Centre, was changed to Nihon Keibi Hosho Printing Centre in order to keep the name Nihon Keibi Hosho in use and thus prevent any other company from assuming it.

At about this same time the government's Administrative Management Agency created a new and separate industrial category for the security industry. Some 850 security corporations were engaged in security activities. Nihon Keibi Hosho was the leader among these but as SECOM its mission was expanding beyond the range of this classification.

When SECOM was officially adopted as the company name Iida felt there was a need for new songs that would more accurately reflect recent and future corporate activities. Keisuke Yamakawa, a poet, and Katsuhisa Hattori, a musician, were commissioned to write the words and music, respectively, for a new corporate song. As well, submissions of original *aishoka* or 'favourite songs' were sought from among SECOM's employees.

Two were selected to be sung on a variety of special occasions. The new company song and the *aishoka* were sung for the first time on 7 July 1984, the day of the *Tanabata* festival and the company's twenty-second anniversary. The texts of the three songs follow.

SECOM Company Song

The blowing wind from tomorrow's ocean,
Is dazzling and new.
The passion of the soul looking into the distance,
Makes the dawn refreshing.
Hoist the sails of hope,
Set forth and travel farther than anyone else.
Build, build, you and I.
SECOM makes a scintillating future.

Sky of cobalt blue,
Its expansiveness and depth is infinite.
Longing for beautiful birds,
Youth is still brilliant.

Spread the wings of freedom,
Take off to the expansiveness of the unknown.
Aim, aim, you and I.
SECOM aims for the distant light.

Footsteps to which hesitation is unknown,
Walk forward across the ages.
One person, one frontier,
All are challengers with compassion.
Chase your ideal as though it were a ball,
Form a scrummage of hearts.
Spin, spin, you and I.
SECOM spins the prosperous earth.

From These Hands

On the palms you extend,
I place my passion.
Brightening the heart of promise,
SECOM's sphere expands afar.
Until friends around the world,
Exchange the smile of peace.
SECOM, SECOM
Let's go hand in hand with passion.

From the earth on which you walk,
Dreams will emerge and flower.
Shining colourfully,
SECOM's sphere expands richly.
With eyes shining like the morning sun,
You can see into the future.
SECOM, SECOM
The door of tomorrow makes wishes come true.

Your singing voice sonorous and clear,
Penetrates to the end of the sky.
The melody of hope takes wings and flies,
SECOM's sphere expands brightly.
As the song leaves the singer's lips,
In order to kindle love in society.
SECOM, SECOM
Paint in hearts an eternal mark.

If I Had Wings

If I had wings,
Could I see tomorrow's happiness?
When people know peace,
Everyone becomes a bird on the wing.
SECOM you invite dreams.
SECOM you are a landmark.

On the long and treacherous wilderness road,
Is the travelling soul lonely?
When people light their hearts with hope,
They will surely become an uplifting breeze.
SECOM even when the night is dark,
SECOM you stand beside me.

Even the floating clouds seen through the window,
Change with the passing seasons.
Unchanging eternal love,
Always desired by people.
SECOM you advance tirelessly.
SECOM you are my friend.

10 Internationalization

FOREIGN EXPANSION

Security is a ubiquitous human/societal need. Based on this assumption Iida believed that Nihon Keibi Hosho's security services could be marketed internationally. He felt in the mid-seventies that the time was ripe. With the growing affluence of the developed and developing countries more money was being spent on the services which exporting countries such as Japan offered. Improvements in financial circumstances which have brought wealth and comfort have stimulated a dramatic increase in the growth of service industries. Iida was cautious, however, about expansion into foreign countries as he had unpleasant memories of his dealings with Erik Philip Sorensen in the early years of Nihon Keibi Hosho's development.

The services Iida wished to introduce into foreign markets had been tested in the domestic arena and he was confident of their excellence. To maintain quality standards, well thought-out standardization of various facets of operational procedures is essential. Nihon Keibi Hosho had devoted considerable energy to the development of detailed operations manuals from its early days. With well-honed security service operational know-how the corporation was ready to expand its business frontiers beyond national boundaries.

However helpful the clear delineation of procedures might be they would be insufficient on their own to form the basis for expansion. Though Iida felt there was widespread international need for his company's security services, he knew that the cultural, social, and political context of each country was unique and demanded respect. Accordingly, he proceeded cautiously. In the early nineties, SECOM operated in four countries outside of Japan: Taiwan, the United States, Korea, and Thailand, and at the time of publication the company had just expanded its operations into Malaysia and England.

One facet of Iida's preparation for foreign expansion was the recruitment of employees experienced in doing business in the international arena. The primary source for such personnel was trading corporations. Such employees have been instrumental in SECOM's expansion into foreign countries, particularly in Southeast Asia where Iida has used them to ground businesses in Taiwan, Korea and Thailand and to head up the Japanese side of these operations. Today, all of SECOM's involvement is orchestrated by the company's international division under the direction of Yasutaka Sugimachi and Hitoshi Wada.

In 1983, six months before the company name was officially changed to SECOM, Iida decided to issue Euro-dollar convertible bonds valued at forty million US dollars. Convertible bonds carry a lower interest cost than straight debt, making them advantageous to the company. From the investor's point of view these bonds offer both the upside potential of common stock and the downside protection of a bond.[1] Iida and his associates travelled to London, Paris, and Zurich to promote the bond issue in financial circles. When the bonds, registered on the Luxembourg Stock Exchange, were issued they sold out almost immediately. The following year eighty million dollars worth of convertible bonds were issued and, again, were purchased rapidly.

The internationalization of service industries is seldom without difficulty. Japan's Economic Planning Agency recently identified eight significant obstacles to the entry of Japanese service industries into overseas markets and to the entry of foreign service industries into Japan. In both cases the primary barrier is the complexity of the administrative and legal procedures that have to be completed before business can begin. SECOM has minimized this problem through heavy reliance on local personnel.

Today SECOM's overseas employees number 7,000 and handle 90,000 contracts accounting for 15 per cent of the company's total sales in 1989. Iida's strategy for foreign expansion, beginning in 1978, has been through mergers and acquisitions in the United States and joint ventures in Southeast Asia, always relying heavily on local human resources, production capabilities, and materials. In the United States, where security services are well established, local personnel make all important decisions in accordance with Iida's view of international business. In joint ventures in Southeast Asian countries personnel are sent from Japan to help launch the new company through the provision of 'know-how', although company presidents are local. While such activities will be taken over eventually by local staff – since the type of business is new to the society – it is felt that the presence of Japanese expertise is beneficial in the early stages to foster an understanding of the social responsibility of the security business.

A 1989 *Nihon Keizai Shimbun* survey indicated that eighty-two of the one hundred presidents of major Japanese companies surveyed were looking for or considering mergers with and acquisitions of domestic or foreign firms as a means of diversifying and/or internationalizing their operations. The acquisition of American firms, in particular, was increasing with 167 takeovers in 1988 compared with 120 in 1987. Iida was in the forefront of this trend, especially in the service industry.

Iida points out that mergers with and acquisitions of domestic firms in Japan are difficult due to the make-up of the management group. Many senior executives in Japanese corporations hold their positions, not because of superior management capabilities, but for other reasons such as seniority. Iida feels that after a merger, if a competent manager is sent in, he will meet with tremendous resistance, much more than in an American company.

By approaching internationalization through merger and acquisition, Iida buys time. To establish a company from scratch in a foreign country and gain a respectable share of the market may take several years. A merger or acquisition allows the company to acquire what Iida feels is one of the most important assets – time. Iida believes that it is essential to assure mutual understanding and trust, especially among the senior executives, when one embarks on a merger or acquisition. Through the meshing of capital, managerial skills, and hardware, the merger will benefit both sides. One-sided benefit is indicative simply of excessive possessiveness. Iida's first experience with a merger was a means through which he could acquire essential skills through first-hand experience:

> My experiences with mergers and acquisitions are not numerous but I have learned that to succeed one has to have a fundamental corporate philosophy and conceptual business design with regard to the future direction.
>
> I started my business, new to society at the time, with the lofty intention of creating an ideal security firm. However, some of the corporations I acquired did not operate with similarly high ideals. Sometimes, after a merger or acquisition I would encounter situations which would make me wonder aloud how the foreign company had survived as long as it had with such weak business practices. Of course, I could not singlehandedly correct the situation. It was only through patience and communication that

staff attitude and morale improved, eventually resulting in corporate advancement. When an acquisition is contemplated it is the quality of the human resources of the company I wish to acquire that is the deciding factor.

AUSTRALIA

Nihon Keibi Hosho's first attempt at internationalization was in conjunction with Brambles Security Group, a component of Brambles Industries Limited, an Australian-based firm providing services in Australia, the United Kingdom, Europe, Canada, and New Zealand. Brambles' activities are diverse and today its operations include industrial plant and crane services; transportation services including sea and air freight and storage; waste collection and disposal; underwater engineering and supply ship servicing for offshore natural resources exploration; business records management and storage; armoured car security transportation; and the worldwide transportation of bullion, precious metals, valuables and documents. Toda recalls:

> In 1977 Brambles personnel visited Japan and approached Nihon Keibi Hosho about the possibility of a joint venture which would have brought about the establishment of a security firm in Sydney.

Iida and Toda believed that with a well-established partner like Brambles, they could succeed in such an undertaking. Together, they visited Australia and concluded an agreement with Brambles' president, Warwic J. Holcroft. Following approval by the Australian government, the joint venture SECOM Australia Limited was established in February 1978. The new corporation's initial undertaking was the conduct of a market survey and the development of a business plan. Says Toda: 'We were confident of our ability to create a first rate security firm but Brambles was less so.'

There were some important differences in the approach of the two companies. Among other things, Brambles wanted to sell security equipment to clients outright while Iida preferred to provide rental/maintenance packages similar to those which Nihon Keibi Hosho offered to its Japanese clients.

Because these differences had not been resolved after three and a half years and neither corporation was fully committed to the joint venture, the decision was made to put the project on hold. Over the intervening years, Iida has gained experience in corporate management in foreign settings. His interest in Australia remains, but in 1990 he entered into an agreement to terminate the tentative Brambles/ SECOM contract, preferring to find a new avenue into the Australian market.

SAUDI ARABIA

In 1978, while Nihon Keibi Hosho and Brambles were negotiating a joint venture, Nihon Keibi Hosho was approached by Chiyoda Chemical Engineering and Construction Company Limited (today Chiyoda Corporation) to provide security at the Ryad Oil Refinery construction site in Saudi Arabia. UOP Incorporated, entrusted with the design and management of the Saudi Arabian Ministry of Petroleum and Minerals' Petromin Project, contracted with Chiyoda Chemical Engineering to construct the industrial complex on a cost-plus, turn-key basis, that is, to construct the plant and prepare it for operation before turning it over to the Saudi government.

Chiyoda Corporation, associated with the Mitsubishi group, is a major Japanese plant engineering company specializing in oil refinery and petrochemical plants. Concerned with the security of the construction site which was located close to the existing oil refinery and with building security at the new facility Chiyoda Corporation turned to Nihon Keibi Hosho, by this time well-known throughout Japan for the provision of security service.

Over the course of the three-year contract (April 1978 to March 1981) Nihon Keibi Hosho built, installed, and operated a localized entry control system and a theft and fire security system for the construction site. Workers from a dozen countries were participating in the project, including 5,000 locals and 1,000 Japanese. Their entry into and departure from specified areas were controlled by magnetic card. The system could identify the location and number of people remaining within any specified area of the complex. The fire and theft security system protected the construction site warehouses.

Norimitsu Matsuzaki and Tomokatsu Kajiwara, Chiyoda's administrative field office managers at the site, point out that despite the sophistication of the security system problems still existed. At peak usage times (particularly starting time and quitting time) the number of workers flooding through each of the ten gates was too great for the security system to process quickly and massive bottlenecks were commonplace. In addition, workers had endless problems with lost, damaged, or forgotten identity cards and cards had to be reissued constantly.

In addition to the electronic security system, Nihon Keibi Hosho also provided round-the-clock, on-site English-speaking security guards at the Ryad site. Says Matsuzaki:

> The guards were efficient and precise in carrying out their duties. They had been well trained for the work by Nihon Keibi Hosho and, in addition, brought to the job valuable experience gained from guard duty at oil refineries and nuclear stations in Japan. The way in which they were to perform their duties was spelled out in minute detail in their company manual and they followed it to the letter.
>
> This sometimes caused problems as the theory did not always translate into the best way to accomplish a particular task given the specific circumstances of the Ryad site. I remember Mr Iida visiting the site briefly to encourage his men and to work out any problems there might have been. We completed the project without any major incident.

With one foot in the door, Iida and Toda approached the Saudi Arabian Ministries of Defence and the Interior about the possibility of Nihon Keibi Hosho penetrating the Saudi market directly, possibly through a joint venture, and were favourably received. They met with government officials and businessmen many of whom – as members of the nation's elite – had been educated abroad and who had no difficulty grasping the concept of an on-line computerized security system.

With the talks proceeding smoothly and the likelihood of success high, Iida and Toda turned their attention to the practicalities of how the system could be introduced to Saudi urban centres. They soon realized that the Saudis were without a spatial identification system that permitted either the precise identification of points on a line (street), as in North American cities, or the identification of ever smaller spatial units as in Japanese urban centres. The Saudis had no address system as we know it.

In the traditional Saudi oasis culture, locations are identified by their proximity to geographical features such as dunes, hills, and *wadi* (dry rivers) which have been given names. Today's urban centres such as Ryad and Jiddah have emerged from oasis towns and still follow this tradition, although the landscape markers are now constructed features such as palaces and mosques rather than natural features.[2]

Saudi Arabia is modernizing rapidly and today the main arteries of urban areas are named. Twelve years ago, however, when Nihon Keibi Hosho was contemplating expansion into the country, the lack of a system of precise spatial identification proved a major stumbling block to the identification of the location of emergency signal emissions, making it difficult to design and operate a centralized security system. Perceiving this to be an obstacle too costly to overcome, given the technology of the time, Iida and Toda called a halt to the negotiations.

TAIWAN

In the Asian arena Iida's approach to internationalization has been characterized by collaboration of effort with foreign entrepreneurs. Eager to expand the security business into Asia Iida looked to Taiwan for a suitable partner with whom to launch a joint venture. He had contemplated such a move as early as 1976, realizing that the country's ongoing industrialization, the recent high growth of its service sector, and the fact that its population density was among the highest in the world, made it a prime target for the introduction of a security service.

Because Taiwan was not universally recognized as a nation, Iida understood that its citizens feared that its currency could be devalued at any time on the world monetary market. Thus many wealthy Taiwanese with connections to the outside world kept their fortunes abroad. Those who did retain significant wealth in the homeland often converted it to American dollars, jewellery, and gold and, not trusting the banks, kept it with them in their houses, an easy target for thieves.

While burglary accompanied by violence or even the threat of violence to the victim carried the death penalty, 'non-violent' burglary would result in an offender being imprisoned for only three months. Consequently, burglars were abundant in Taiwan. In addition to the likelihood of being robbed, Taiwanese faced a considerable threat to their property from fire.

The late Noboru Goto, head of the Tokyu Group, suggested to Iida that if he was seeking a Taiwanese partner for a joint venture he should look to the nation's major companies. In particular, he recommended Taiwan Cement. Iida sent an employee to Taiwan to do the ground work. He was put in touch with the late Naosuke Tomita, Japanese military adviser to the Taiwanese government, who in turn suggested Lin Teng as a suitable partner. In 1977 Iida had the opportunity to visit Taiwan with Goto who provided him with an introduction to Lin Teng of Taiwan Cement.

Lin Teng had gone to Japan in about 1940 to study business administration at Osaka University. While there he acquired knowledge about the slate industry and on his return to Taiwan successfully launched a slate business. After the war, when the

Japanese withdrew from Taiwan and the four major government-owned corporations were privatized, Lin Teng became a senior executive with Taiwan Cement Corporation. Later, through outright ownership and through investment, he expanded his business into real estate, television broadcasting, airline service, commercial rental properties, construction, trading companies, securities, hotels, and foreign investment including Saudi Arabia's national cement company. He also entered into joint ventures with Meiji Milk Products Company of Japan and Scotties of the United States. Today, Lin Teng's conglomerate, Koushan Enterprises, consists of some twenty companies.

Lin Teng knew about the security industry in Japan and felt that a similar industry in Taiwan could provide a useful service and a good business opportunity. While Taiwanese law required that bank security be provided by the police, other corporations maintained in-house security. They sought guards from among retired military personnel, many of whom had come into the country with Chiang Kai-shek. There were no companies in Taiwan providing security services either through on-site guards or electronic monitoring systems.

Lin Teng was favourable to Iida's business proposition and offered to involve family members in the new venture. Because the establishment of a security service joint venture was prohibited under Taiwanese law, it was agreed that Lin Teng would first register a new company with a security focus and afterwards work at developing it into a partnership with Nihon Keibi Hosho.

In November 1977 Lin Teng initiated Chung Hsing Security Communication Co. Ltd with the intention of it becoming Taiwan's first security firm. It would be another year, however, before the government would grant permission for the new company to act as a security firm and for Nihon Keibi Hosho to invest in it.

The government perceived the existence of security firms as a potential threat to the military. All important positions within the government, the military, and the police were held by Chinese who had come to Taiwan from the mainland with Chiang Kai-shek and who maintained that Taiwan was the proper seat of the Chinese government. Business was dominated by native Taiwanese.

Many native Taiwanese did not necessarily share the government's view and some argued that Taiwan should become an independent nation. Chiang Kai-shek's nationalist government held its political stance with the backing of the military and police force which existed primarily to fulfil the wishes of the government rather than to assure public safety. The government's policies were repressive. No gatherings of any kind were allowed. The nation's communication channels were under the control of the military who were against the idea of private sector use of communication lines for security purposes.

For these reasons Chung Hsing Security Communication could not immediately enter into the security business. Initially, the company was involved mainly in the sale of Nihon Keibi Hosho's fire extinguishers. Despite the fact that the company sold 120 halon gas extinguishers to the National Palace Museum in Taipei, the high import duties necessitated the company taking a loss on the sales. Before the company could function as a security firm approval had to be obtained from the military, the police, the Ministry of Trade and Industry, and the Ministry of Communications.

To increase the likelihood that approval would be forthcoming, Lin Teng, who was the chairman of Chung Hsing Security Communication, recruited the assistance of Hsiao Cheng-chih. A military officer who had come to Taiwan with Chiang Kai-shek, Hsiao Cheng-chih had a close relationship with Chiang Kai-shek's son and could exert considerable influence on the political scene. Lin Teng made him vice-chairman of Chung Hsing Security Communication. The president was Lin Teng's second son, Lin Roscow, a graduate of Taipei Institute of Technology who had studied managerial engineering for two years at the Science University of Tokyo. He recalls:

> When I returned to Taiwan I worked in areas of my father's business empire unrelated to my formal study. First I managed a 120-room hotel, then a travel agency, then a construction company. Now, in addition to Chung Hsing Security Communication, I'm involved with securities.

To complete the staff of the new company Lin Teng hired two Taiwanese secretaries and Iida sent three Nihon Keibi Hosho personnel to Taiwan. One of these was an engineer, one was in sales, and the third, Fumio Abo, was to be the coordinator of the project and the liaison between Nihon Keibi Hosho and Chung Hsing Security Communication.

After graduating from Gakushuin University in 1948, Abo worked for the Daiichi Trading Company, the precursor of the Mitsui Trading Company, for eight years before joining the Nippon Signal Company. In 1976 Abo joined Nihon Keibi Hosho and spent a year in the United States coordinating the technical tie-up between Nihon Keibi Hosho and the Codata Corporation. For the next nine years he served as the liaison with Chung Hsing Security Communication while simultaneously maintaining his position with Nihon Keibi Hosho, first as director of sales for fire extinguishing equipment and later as the managing director of all domestic sales.

Abo's first task, with the help of his Japanese colleagues, was to ensure that Hsiao Cheng-chih was well acquainted with the relevant aspects of the Japanese situation regarding private security systems and private security law. Communication was difficult as Hsiao spoke only Mandarin. Abo had to rely on Lin Roscow, who spoke Taiwanese, Mandarin, Cantonese, and Japanese, to act as interpreter. When Hsiao finally had a clear picture of the situation he spent six months visiting various government departments and politicians. During this six-month interval government restrictions prohibited Abo from reporting any information whatsoever to Nihon Keibi Hosho concerning the progress of Hsiao's negotiations. Towards the end of this period Iida's Taiwanese advisers sensed that government opinion favoured Chung Hsing Security Communication and recommended that Iida come to Taiwan to meet and socialize with government officials. Iida organized a party for all the high officials involved with Chung Hsing Security Communication's case. Recalling that time, Iida says:

> Before the party started the chairman of a Taiwanese economic association took me aside and told me that Chung Hsing Security Communication's permit was almost certain to be issued, especially if I took care to drink to the hilt, exchanging glasses of brandy with everyone there. I followed his advice that night and the next day golfed with some of the officials. Almost as soon as I got into my car at the airport on my return to Japan the phone rang. Chung Hsing Security Communication had its permit! Drinking had sealed the deal.

After Chung Hsing Security Communication was registered as a domestic company it increased its shares and Nihon Keibi Hosho was permitted to purchase 30 per cent of them. The Taiwanese Incorporation Act specifies that 10 per cent of a company's shares must be owned by its employees. The remaining 60 per cent of the shares belonged to Lin Teng and his family and friends.

At that time, towards the end of the 1970s, joint ventures between Taiwanese and foreign companies were controlled by strict government regulations. Lacking sufficient foreign currency reserve, the government permitted joint ventures only with foreign manufacturing companies, stipulating that 40 per cent of the manufactured product had to be exported. Nihon Keibi Hosho could not satisfy this criteria so instead of a joint venture the company entered into a technical assistance contract with Chung Hsing Security Communication through which it would provide equipment and management expertise and in return receive royalties.

The interpretation of the meaning of this type of agreement created problems. Hsiao Cheng-chih, being a military man, understood 'technical assistance' to mean the provision of equipment by foreigners without charge, as in the US provision of weapons and ammunition to Taiwan under Chiang Kai-shek. In the process of negotiating the contract Iida met with Hsiao who insisted that Nihon Keibi Hosho, a large, well-established company, should provide all equipment free to Chung Hsing Security Communication, a small, newly emerging corporation. Says Abo, who was present at the meeting:

> Mr Iida pounded the table and pronounced the proposition outrageous. He argued that the contract was between private companies, not nations, and that Nihon Keibi Hosho had no intention of providing its equipment free to the Taiwanese. He acknowledged that the equipment would be provided at an appropriate discount but that it would carry a price tag fair to both sides. As a result of this direct confrontation, respect and a strong friendship developed between Mr Iida and Mr Hsiao.

The equipment was exported from Japan to Taiwan but a 45 per cent import tax was placed on the f.o.b. price making the service very expensive for Taiwanese customers to purchase. Lin Roscow recalls:

> Business was slow. Our potential customers had no clear idea of what electronic security was and I didn't understand the concept thoroughly either. Nevertheless, we managed to negotiate twenty-two contracts the first year, although ten of these were to members of Koushan Enterprises that my father heads.

As in Japan, equipment was leased rather than sold to clients outright. When the paucity of contracts resulted in insufficient capital to cover operating costs Chung Hsing Security Communication sought a bank loan. Because the company was new, in debt, and without assets to serve as collateral the loan was not forthcoming. Privately wealthy, Lin Teng offered to act as a guarantor but this was rejected also. The problem was finally solved when a bank in Taiwan and Sumitomo Bank in Japan entered into an agreement whereby Nihon Keibi Hosho would stand as the guarantor for the letter of credit. A reduced rate of interest was also negotiated – 12 per cent instead of 15 per cent – but increasing sales in order to reduce the expenses incurred per contract remained the crucial issue.

Gradually contracts increased and by 1982 numbered 500. While per contract operating costs decreased as the business expanded, the import taxes levied against the Japanese hardware remained a major financial hurdle. Importing the SP Alarm System hardware presented other problems as well. Imported hardware was often held up in customs and unavailable for implementation when the new contracts were signed. Similarly, if a breakdown occurred and new parts were required they were not always immediately available.

Taiwan had a sufficiently high technological capability to warrant local production of the electronic components of the SP Alarm System but the initial contract volume did not justify this activity. As contracts increased through the early 1980s the decision was made to launch a Taiwanese production system. Lin Teng established a factory hoping to reduce costs at least by the amount of import duties. By 1985, 75 per cent of the hardware was manufactured in Taiwan. In this way production costs were minimized and security service costs were no longer affected by price fluctuations resulting from a variable foreign exchange rate.

From the outset, quality control has been a major concern for both Iida and Lin. Faulty equipment, particulary sensors, would cost the company dearly. Locally made products from all of SECOM's overseas ventures are tested in Tokyo before being introduced to the market and, thereafter, on a spot-check basis. Because of the high quality standards, even when produced in Taiwan, Lin Roscow believes SECOM alarms and sensors are five times more expensive than others produced domestically. But in the long run the high standards pay for themselves by minimizing system malfunctions.

Equipment produced in Japan could not be used in Taiwan without adaptation. The hardware for export had to be produced in accordance with local specifications. Since the production site moved from Japan to Taiwan, efficient, economical copying and adaptation of the Japanese products has been a major focus of attention for the Taiwanese. Chung Hsing Security Communication recently created a small research and development centre. The six staff members are engaged in research related to the improvement of the present SP Alarm System. In the future their mandate will be broadened to include investigation into new business opportunities.

As the number of contracts increased so did the number of employees. In the first one or two years after Chung Hsing Security Communication's establishment, security guard personnel hired included many former military police from Chiang Kai-shek's forces. They were among the two million Chinese who were driven out of

China when Chiang Kai-shek's nationalist government was overthrown in 1949. Forced to leave their families behind, they joined the six million inhabitants of Taiwan and were refused contact of any kind with mainland China. Taking care of them as they aged – many of them were single – became a government concern.

In 1954 the government established a committee to oversee the medical needs, continuing education, and reemployment of these emigrés.[3] The emergence of Chung Hsing Security Communication and, later, several other security firms, as an employer of retired military personnel was welcomed.

These emigrés, however, were aging rapidly. Chung Hsing Security Communication needed younger men and more of them. The company sought out young university and high school graduates who had completed the compulsory military training of two and three years respectively. Recruitment was not difficult but retaining employees presented a challenge; many employees tended to leave the company after a few years. Abo reports:

> In Taiwan it is virtually impossible to climb the corporate ladder to the executive ranks unless one is related to the company's owner. As a result young people often work for a company only for a few years, until they have enough money to go into business for themselves.

Lin Roscow corroborates this.

> In Taiwan we often say that one out of every three working people is a company president. A common pattern is for the husband to work for the government or a large corporation and for the wife to manage a family business.

There are numerous small corporations in Taiwan. The ratio of companies to population is one of the largest in the world. In order to decrease the rate of employee turnover Lin Roscow and Hsiao developed and implemented an incentive plan. Sales personnel are awarded a cash bonus each time a new contract is signed. Sometimes the monthly earnings of a salesperson equals that of Lin Roscow. Likewise, guards are also eligible for cash bonuses. They receive a bonus equal to approximately four times their monthly salary each time they apprehend an intruder.

The most common criminal offence in Taiwan is burglary. In 1986, for example, out of some 93,000 criminal convictions, 52 per cent were for burglary, by far the largest type of offence. The next largest category was gambling but this accounted for only 9 per cent of the total convictions. Armed robbery and any type of resistance or violence on the part of the burglar carried the death penalty, so burglars almost always surrendered to the guards without a struggle. With little risk to personal safety and the promise of a large monetary reward guards would speed to the scene when the alarm sounded, always beating the police by several minutes.

Initially, despite the high burglary rate, business in Taiwan was slow. Over the intervening decade, however, the public, especially the high income group, has become more security conscious. When the number of contracts reached 3,000 the business began to take off and profitability improved. By 1989, twelve years after establishment, contracts numbered 14,000. Although the geographical/historical context differs, when Nihon Keibi Hosho became listed on the Tokyo Stock Market in 1974, twelve years after its establishment, its contracts numbered 12,000. Today, with 20,000 customers, the Taiwan company's economic performance remains good.

While Chung Hsing Security Communication is not a joint venture, Iida continues to send personnel from Japan to assist not only in the technical area but also in sales. One of these men, Hirofumi Onodera, after graduating from the Muroran Institute of Technology, joined the company in 1977. After gaining experience in sales engineering Onodera was assigned to the company's overseas project division in 1980. His first foreign posting was to Taiwan in 1981 as a sales engineer. Onodera became vice-president of the company in 1989, at Lin Roscow's request.

Onodera's inability to speak either Taiwanese or Mandarin restricted his activities both in providing advice and in selling. Within Chung Hsing Security Communication he could communicate only with those who spoke Japanese, and in his search for new customers he limited himself to Japanese nationals and to those elderly executives of large corporations who spoke Japanese, a legacy of the Japanese occupation.

Onodera considered that his mission in Taiwan was to improve the performance of Chung Hsing Security Communication. To do this he felt it was necessary to become, as much as possible, a 'local person' so that he could appreciate the situation from the Taiwanese point of view. To accomplish this, acquisition of the language was essential and he worked hard to develop his communication skills.

By the time I joined SECOM it was a well-established company. Through my involvement with Chung Hsing Security Communication I was able to experience some of the same type of challenges and excitement associated with clearing a path and pushing back the frontier, that my more senior colleagues had experienced in the early days of Nihon Keibi Hosho.

In Taiwanese the words for 'security' and 'insurance' sound very similar when pronounced. They share a common first character so even the written words are easily confused. This fact coupled with the widespread lack of familiarity with security systems creates confusion. Taiwanese salesmen sometimes fail to clarify the difference when dealing with potential customers.

Because of the frequency of theft and the difficulty of proving what was actually taken, property insurance does not exist in Taiwan. Chung Hsing Security Communication, like SECOM, is only responsible for client compensation when a 'successful' burglary is the result of a system malfunction.

In Japan proof of the value of stolen items must be provided and must be approved by both the police and the insurance company before SECOM pays compensation. In Taiwan, however, many valuable items are obtained on the black market, making difficult substantiation of their worth for purposes of compensation. Especially at the outset of Chung Hsing Security Communication's development customers would sometimes take advantage of this situation.

As Chung Hsing Security Communication grew the company confronted some of the same problems that had plagued Nihon Keibi Hosho in its developmental stages. In 1986 the first and only employee theft occurred, damaging morale and causing widespread concern over how a recurrence could be avoided. Iida, however, viewed the event as a rite of passage marking the company's entry into adulthood. While Lin Roscow acknowledged that such an event was virtually unavoidable as the number of security guards increased, he decided to establish a human development centre in Tanshui near Taipei. Used primarily by Chung Hsing Security Communication, other members of the Koushan Group have access to the facility when it is not required by the security firm.

New employee training and follow-up seminars take place at the human development centre. Emphasis is placed on the conceptual and on the mental state of employees, especially on personal integrity and social responsibility. Technical aspects of the job are dealt with primarily in on-the-job training.

Unlike in Japan, where there is only one starting date a year for new employees nationwide, in Taiwan employees are hired as needed and may start anytime. This makes difficult the coordination of the new employee programmes at the human development centre. As the company grows, however, and greater numbers of new employees are hired, both in response to the growth and to replace staff who are leaving or retiring, the multiple starting times may prove to be convenient as small groups of employees can receive training at the centre soon after they are hired.

Today Chung Hsing Security Communication's 1,100 employees work at the company's headquarters, thirty-seven branch offices, and three control centres in the country's three major cities. Taipei, Kaohsiung, and Taichung, Taiwan's largest cities and the cities with the highest crime rate, are situated on the west side of the island where most of the company's 22,000 clients are located. Of these, half are serviced by private lines while the remainder rely on public telephone lines. Chung Hsing Security Communication would like to expand its service area to the eastern side of the island but the high mountain ranges that run the length of the country and account for 70 per cent of its area interfere with reception on the east coast.[4]

Thus far the majority of Chung Hsing Security Communication's clients are institutional users, corporate offices, and merchants whose residences are frequently within their business premises. Banks, schools, museums, factories, and in the south, orchid and other flower growers gradually are beginning to use Chung Hsing Security Communication's services; these are the target customers of the future.

The nature of crime in Taiwan appears to be changing with gang murder and armed robbery on the increase and youth crime increasing.[5] Taiwan's recent modernization and success in international trade have resulted in rapid economic progress and a significant increase in the standard of living of the population at large. In the last five years the per capita income has doubled, real estate and stocks have sky-rocketed, and the number of imported goods has increased with the growing consumer purchasing power and trade liberalization. Hand in hand with the improved economy has come an increase in the crime rate. A recent report by the Directorate-General of Budget, Accounting, and Statistics indicates that 43 per cent of the companies surveyed had been threatened with extortion or kidnapping while 34 per cent had been robbed or burgled.[6]

Such well-known foreign products as Mercedes Benz automobiles, Rolex watches, and Celine handbags are popular among the wealthy and sell well in Taiwan. Last year, however, Mercedes Benz sales dropped by 20 per cent. According to Lin Roscow:

Mercedes Benz have become the mark of the rich, making it easy for thieves to identify potential victims to mug or even kidnap. I try to protect myself by using my Mercedes only in the daytime and switching to a much less expensive, more ordinary car at night when most incidents occur.

Taiwanese have started to use credit cards but, generally, they still carry large amounts of cash on their persons. According to the results of a recent Chung Hsing Security Communication survey, women are likely to have in their handbags at any given time the equivalent of about half the monthly household income.

The police encourage the use of bank vaults for the storage of valuables but custom is hard to change. They also recommend that if the entire family is going out that they do so in a staggered fashion rather than all at once to give the impression that the house is still occupied. They also warn individuals to beware of salesmen, delivery men, and even policemen calling at the door. They could be disguised thieves prepared to force their way in to the residence once the door is open.[7]

Against this background many security firms have sprung up in Taiwan, some of which have dubious reputations. As in Japan, laws regulating the establishment of such firms are in the process of being developed. The absence of a security association in Taiwan makes it difficult to ascertain precise statistics on the security industry but there are at least ten firms offering electronic security and among them Chung Hsing Security Communication has captured over 60 per cent of the market. There are some thirty firms offering human-based security only and numerous firms which sell burglar alarms and related equipment.

Although the company sold fire extinguishers when it first began, Chung Hsing Security Communication has not yet entered into the fire security business. Taiwan lags far behind in fire prevention and detection measures. The hardware infrastructure is incomplete and not dependable. Alarms designed to be connected to the fire station often are not. Laws governing fire prevention and detection exist but are seldom followed. Chung Hsing Security Communication is reluctant to venture into fire detection services until the supporting social infrastructure is stronger. With regard to the extension of basic services Chung Hsing Security Communication is contemplating VAN service and personal medical services.

SOUTH KOREA

SECOM's entry into the South Korean market was accomplished via a joint venture with Samsung Company Limited, South Korea's oldest and biggest conglomerate. Today its sales account for 13 per cent of South Korea's gross national product. Founded in 1938 by the late Lee Byung-chull as a trading company, during the fifties Samsung expanded into manufacturing becoming the nation's first producer of sugar, wool and textiles. Within a few years Lee also controlled three South Korean banks. In the sixties Samsung expanded into paper and publishing, consumer electronics, fertilizers, retailing, and life insurance, followed, in the next two decades, by moves into petrochemicals, hotels, construction, and industrial electronics.

The vast number of Samsung-owned facilities requiring security personnel coupled with rising labour costs caused Lee Byung-chull, Samsung's chairman, to consider the introduction of a modernization plan in which machine security would largely replace human-based security. Entering into the provision of this type of security would provide yet another avenue of expansion for the conglomerate. Throughout Samsung's development, especially in its advance into electronics, Lee had looked to Japan as a role model.

In considering expansion into the security field he did so once again. He instructed Samsung Japan, the company's Tokyo office, to investigate and put forth a recommendation regarding an appropriate Japanese security firm with which the Korean corporation could enter into a cooperative business venture. Samsung Japan recommended Nihon Keibi Hosho. In 1979, with a Keio University Human Resources professor acting as a go-between, Iida and Lee met and agreed to involve their respective companies in a joint venture.

Iida appointed Yuji Nishimura to head a committee, formed in April 1980, to launch the joint venture. Prior to joining Nihon Keibi Hosho in 1976 Nishimura had worked for twenty years with Ataka Trading Corporation. Before excessive investment in the petrochemical industry in eastern Canada resulted in severe financial difficulties that led to Ataka's absorption by C. Itoh & Company, Nishimura had gained considerable experience in international business dealings. After joining Nihon Keibi Hosho he was, for three years, director of the company's Yokohama branch. With three hundred employees under him he was responsible for the entire Kanagawa region, giving him a broad understanding of the security operation.

The mandate of the committee Nishimura headed was to plot a course of action and lay the foundation for the new company to be established as a joint venture in Korea. The first challenge confronting the committee and its Korean counterpart was the negotiation of a contract suitable to both sides. Underlying the discussions was a deep-rooted distrust of the Japanese on the part of the Koreans.

Japan has a long history of invasion and domination of the Korean nation dating back to the late sixteenth century when Hideyoshi Toyotomi invaded the country and was driven back only after six years of devastation. During the Sino-Japanese War (1894–5) and the Russo-Japanese War (1904–5) Japanese troops moved through Korea to attack Manchuria. These troops were never withdrawn and in 1910 Japan formally annexed Korea. Although Japanese rule, which continued until 1945, effected some contribution to economic and social development, it was harsh and exploitive. The Samsung executives, mostly in their forties, with whom Nishimura's committee was negotiating, had grown up in a profoundly anti-Japanese environment. Says Nishimura:

> It was hard work to draw up the contract clause by clause. Despite the fact that Korea looks to Japan for an economic role model, because of their inbred distrust of the Japanese the men with whom we were dealing were always on their guard against possible trickery. It was difficult for them to accept anything at face value.
>
> The language barrier contributed to our problems. The Koreans wanted to conduct the negotiations in Korean while we preferred Japanese. The compromise was to use English. Reaching agreement on the wording of each clause was a slow process.
>
> By the summer of 1980 the basic joint venture agreement, which included within it a technical assistance clause, had been drawn up. Despite the fact that we spent so much time formulating a very precise contract now, thanks to the growth of mutual trust and respect, both sides are able to look the other way when slight, unavoidable deviations from the contract occur.

With the basic agreement completed the next task was to secure the permission of the Korean government to establish the company to be known as Korea SECOM. The political and economic climate in Korea was unstable. The president was assassinated in 1980 and strong military control was then imposed on the people. The nation's economy was failing and there was little stability within the country. Under these conditions small- and medium-sized security guard corporations existed. Hearing of the possible joint venture between the giant Samsung and Nihon Keibi Hosho, these smaller companies were voracious in their protest against the formation of a nationwide foreign invested security corporation. Included among them was Korea Security Corporation, a newly licensed, not yet operating machine security company. The company was owned by three investors one of whom was a lawyer and another a politician.

Obtaining a licence to launch Korea SECOM looked as though it would be a time consuming process fraught with difficulties. To circumvent these obstacles and more quickly launch the company, Korea Security Corporation was approached and a tie-up effected between this company and Nihon Keibi Hosho. With one partner in the joint venture now a licensed security firm Korea SECOM could come into being.

The company shares were immediately increased and Samsung bought in. When Korea SECOM was formally established in April 1981 Samsung and Nihon Keibi Hosho each owned 35 per cent of the company with the remaining 30 per cent divided equally among the three owners of Korea Security Corporation which remained a company on paper only. When one of these investors died his shares were divided equally between Samsung and SECOM. Now each of these corporations owns 40 per cent of the company, satisfying Korean foreign investment laws.

Korea SECOM sought the permission of the Korean government to import all necessary hardware from Japan. The joint venture contract stipulated that it was Samsung's responsibility to secure an import licence. The government objected on the grounds that the country's foreign reserve was to be used primarily for the purchase of natural resources and that Korean labour could and should be used for the provision of human-based security. Machine security was not necessary. Eventually, however, negotiations resulted in the granting of a licence. The way was now open to hire employees and launch the business. The president was Korean and Nishimura was installed as vice-president. Four other Nihon Keibi Hosho personnel were sent to Korea SECOM to assist with sales, engineering, and system installation.

Hiring personnel suitable to work as beat engineers and guards was easier than in Japan. Military service of six to eighteen months, depending on education, is compulsory in Korea and there was an abundance of young men who had completed this service looking for jobs. They were well disciplined, well acquainted with the country's geography, competent in the operation of short wave radio and they possessed driver's licences, a particular bonus in a country where cars were scarce and few knew how to drive.

Salesmen, and to some extent beat engineers and guards, were recruited from the large human resource pool within the Samsung conglomerate. Then as now, all hiring for the Samsung group was undertaken centrally by the Human Resource Management Committee. In addition to the in-house employee training provided by all Samsung group companies, Samsung also operates the Youngin Training Centre where comprehensive training is provided for all levels of employees throughout the conglomerate.

As in Japan, the sales strategy employed was to interest financial institutions in SP Alarm System installation, knowing that other organizations would follow their lead. Other early users of the system were corporations within the Samsung group. These corporations also use on-site guards extensively and today half of the company's 1,500 guards are used within the conglomerate.

For many years the military imposed a midnight to 4 a.m. curfew on the nation. Nothing could move during this time. Korea SECOM's guards and beat engineers were licensed to continue to work during the curfew but were frequently apprehended and questioned by military patrols. These patrols were often familiar with the idea of on-site guards as private security corporations were sometimes contracted to provide security at military sites. They were not, however, familiar with machine security so guards and beat engineers were frequently required to show their licences and explain the security system.

Nishimura and the other Japanese personnel worked for the benefit of the Koreans as much as for Nihon Keibi Hosho, striving to understand the Korean point of view and win their trust. Having been invaded many times throughout history and having been at the mercy of foreign rule until recently, the Koreans had developed a strong and forthright will and a straightforward approach.

The Japanese, on the other hand, pride themselves on the ability to feel one another out on an issue without ever confronting it directly. The highest form of interpersonal communication is *haragei*, literally 'the art of the stomach'. The stomach is perceived as the seat of the emotions. One's deepest desires are communicated in the fewest number of words.[8] Among Japanese who share a common history, language, and culture, *haragei* works well. In communicating with foreigners, however, misunderstandings often occur.

Nishimura and his Japanese staff pursued seriously the study of Korean. Nishimura had a private tutor with whom he spent two hours three times a week. According to the contract all reports and documentation were to be in English but this proved difficult as few employees were sufficiently fluent in the language. Initially, when

Nishimura attended meetings with his Korean colleagues he understood little of what had transpired. But gradually as he forced himself to use Korean his comprehension increased. More and more in matters of dispute he found himself siding with the Koreans against Nihon Keibi Hosho. This earned him the admiration and friendship of his co-workers.

For the first three years Korea SECOM carried a heavy debt load. To service even one contract the entire system had to be in place. This included a fully functioning control centre, several beat engineers, and two patrol cars. Contracts were slow to come at first. There was a widespread reluctance to permit strangers access to one's premises both commercial and private. Perseverance and diligence on the part of Korea SECOM employees helped overcome this problem. As the number of contracts increased in Seoul profits were directed to the establishment of branches throughout the country. 'Initially', says Yutaka Wada, sent from Japan to be responsible for technical assistance, 'the Korean employees had difficulty comprehending the relationships among the individual parts that make up the total security system.'

Korea had the technology necessary for production of the system hardware. As the number of contracts increased and the understanding of the hardware on the part of employees deepened, in 1984 local production was contracted to a small, technologically competent company. Although Samsung excels in the manufacture of electronic products it was not economically feasible for the corporation to undertake production of the security hardware. Today almost all of the hardware used in Korea is domestically produced.

By June 1985 contracts numbered about 1,000. With Korea SECOM securely on its feet Nishimura was recalled to Japan and Katsumi Senno sent to Korea in his place. Senno's job was to 'manage' the joint venture from the Japanese side. The running of Korea SECOM was in the hands of the Koreans.

Senno joined Nihon Keibi Hosho in 1978 after twenty-two years with Chori Trading Company during which time he had been stationed in North America, Europe, and the USSR. Realizing the limitations of the trading corporation he sought employment with a company strong in technology and chose Nihon Keibi Hosho. Says Senno:

> In the age of globalization the traditional model of trading company activities had some limitations. Without manufacturing, the product is simply transferred between points and profit

amassed in the homeland. Such a purely Mercantilistic practice, I thought, gave rise to economic friction.

In the early stages of Japan's economic development trading corporations were necessary for the success of manufacturers lacking both financial and human resources. Chori Company was strong in textiles and later machinery. As manufacturing industries improved their financial situation and acquired international experience the justification for the existence of trading corporations lessened.

With the technological level of merchandise on the rise, salesmen were required to have specialized knowledge in order to successfully market the products with which they dealt. Although most trading corporations were diversifying their activities, I thought that in the future technology-based super corporations would dominate the market.

Throughout my career I acquired international negotiation skills dealing with a variety of personalities and nationalities. I wanted to invest my energies in Nihon Keibi Hosho, a technological leader among Japanese security firms.

As clients became more numerous their security needs became more diverse. The original contract drawn up between Samsung and Nihon Keibi Hosho to launch Korea SECOM proved to be too narrow and restrictive to accommodate the variety of situations now encountered routinely by the company. New technology was constantly being developed and integrated into security activities in Korea and the joint venture contract required constant updating to reflect these advances.

Contract revisions alone were not sufficient to propel the company smoothly forward. Senno, like Nishimura before him, threw himself into the task of learning the Korean language and understanding the Korean point of view. But he found that for both the Japanese and the Koreans long standing prejudices were hard to overcome. Added to this was the more recent conflict between the Japanese as the technology providers and the Koreans as the technology recipients and adapters.

In November 1987, two years after Senno was sent to Korea SECOM, Samsung's chairman, Lee Byung-chull, died after a decade-long battle with cancer. His third son, Lee Kun-hee, who succeeded him as chairman, endorsed the joint venture. Nevertheless, there were times of uncertainty for Korea SECOM during the transition phase. Throughout this period, in his conduct of business, Senno acted in accordance with his firm belief that the joint venture would endure.

If I had doubted the permanency of the joint venture I might have hesitated to act positively. This in turn might have created doubt on the part of the Koreans and set in motion a negative cycle.

My belief in Korea SECOM's future grew out of my faith in Mr Iida, acquired from observation of and participation in his business dealings. Mr Iida always follows through on his promises.

The year Lee Byung-chull died Korea experienced seven reported armed robberies and 247 reported unarmed robberies for every 100,000 population. In reality these figures were much higher but the crimes went unreported due either to fear of retaliation on the part of the victim or to unwillingness to disclose the source of one's wealth to the authorities. In any event, crime was on the rise.

In addition to Korea SECOM's first targeted clients – financial institutions including many that were Japanese owned – schools emerged as frequent users of the security service. By 1990 Korea SECOM had 13,000 contracts in the nation's urban areas, with control centres located in Seoul, Inch'on, Teijong, Taegu, Pusan, and Kwanju supported by twenty depots. Many Japanese corporations have factories in Korea's industrial zones located in rural regions. The demand for security is there but Korea SECOM's service network is not yet sufficiently broad to encompass these areas. It is a goal of future expansion to create a more intensive, better integrated nation-wide service. Recognizing the potential business opportunity, Korea SECOM's competitors are increasing in number. While the tally for human-based security services is uncertain, there are at least ten machine-based security services operating throughout the country. Korea SECOM, with approximately 17,000 customers, has captured 80 per cent of the market.

With the hurdles of establishment and the growing pains of the early years behind Korea SECOM, Lee Kun-hee appointed Lee Dong-woo president in 1989. Lee Dong-woo brought to the job a wide range of managerial experience. After graduating from Seoul University in economics and later from military college, Lee spent twenty years in the army, retiring as a colonel in 1974. He worked for a Samsung-owned hospital until 1980 when he moved to Samsung Electronics' semi-conductor and telecommunications department. As a senior manager for nine years, he was involved in research and development, quality control, domestic and international sales, strategic planning, and general control. Says Lee Dong-woo:

As Korea SECOM expands there is increased interaction between Korea and Japan. More and more of our employees are being sent to Japan for training. At the present time we are adopting all aspects of the Japanese business from the technical to the spiritual regardless of the appropriateness to our situation. The Japanese have put so much effort and money into SECOM to make it the success that it is. I think we should learn as much from them as we possibly can. Later, if necessary, we will make adjustments to render Korea SECOM more distinctly 'Korean' and more suitable to the local environment.

Lee Dong-woo speaks excellent Japanese. Japanese language training for executives is provided throughout the Samsung group. Realizing the importance of language fluency for Korea SECOM's staff, Lee promotes the daily ten-minute language broadcast at the company first thing in the morning to which all employees listen. Through periodical tests employees are given the opportunity to have their fluency in Japanese evaluated. Excellence is rewarded with cash bonuses.

Since May 1990 Korea SECOM's vice-president has been Michio Matsuyuki, sent from Japan to replace Katsumi Senno. After graduating from Boei University Matsuyuki was with Japan's self-defence forces for fourteen years, retiring in 1974 from his position as battalion commander. The next year he joined Nihon Keibi Hosho and since then has occupied a number of different managerial positions. Like his predecessors at Korea SECOM he is eager to learn Korean and to widen the path that they have forged.

THE UNITED STATES OF AMERICA

The high crime rate in the United States has been paralleled for some years by an equally high interest in security, much higher than in Southeast Asia. Most Americans either have been the victim of a crime or know someone who has. In 1982, for example, nearly twenty-five million American households – 29 per cent – were victimized by theft or violent crime, and these figures do not include homicide and many unreported incidents.[9] Expansion of municipal police forces is limited by available tax dollars. As well, public trust in the police has been undermined by incidents of corruption. Several thousand private security companies have emerged over the years in the United States to fill the gap in police protection.

Iida recognized the opportunity to market his security services in the

United States in the late seventies. He had been attracted by the size and challenge of this market for some time. In July 1980 he established an office in Los Angeles, staffed by two men from Nihon Keibi Hosho and an American secretary, in order to explore the possibility of entry into the American market.

The opportunity came in February 1981 when he had the chance to acquire the Fresno-based Valley Burglar and Fire Alarm Corporation. This was Nihon Keibi Hosho's first foreign acquisition. The company was acquired through negotiation and most of the staff was retained. Iida stresses the importance in overseas ventures of retaining local personnel as the core of the new company. Throughout SECOM's overseas operations the company presidents are natives of the host country, as are most of the employees.

In each of SECOM's Southeast Asian joint ventures the vice-president is a Japanese national responsible for the management of the Japanese side of the undertaking. In North America the only person-nel sent from Japan are young men whose function is to share technical knowledge and sales know-how in the early stages of the company's development. They are particularly useful in recruiting customers from among Japanese nationals and corporations in the foreign country. As well, they learn from the local operation. In the case of SECOM's American operations this is especially important as the corporation is involved in activities into which SECOM has not yet been able to expand in Japan.

When Nihon Keibi Hosho acquired Valley Alarm, Iida retained the name and installed the son of the former owner as president. Today this company is in the commercial and residential security business. In addition to burglar and fire protection, Valley provides closed circuit television, access control, and industrial monitoring services.

With regard to the hardware that is the backbone of SECOM's security systems, Iida's approach has always been to use locally manufactured components whenever possible. The countries in which SECOM has interests have the technology to accommodate these needs. Costs are reduced with this approach and it contributes to the prosperity of local industry through the provision of employment and technological transfer.

Nihon Keibi Hosho's second American acquisition, again in South-ern California, occurred in October 1982 with the purchase of Westec Security Corporation. Westec had grown to be the largest franchise of Westinghouse Electric Corporation when the franchise owners col-lectively bought the company's patents from Westinghouse along with the right to use the name 'Westec'. The company continued to grow in the provision of commercial and residential security through security patrols and the manufacture and service of security alarm systems.

Nihon Keibi Hosho had been seeking an appropriate avenue for expansion of the security industry in the United States. After considerable deliberation and lengthy negotiations, SECOM purchased Westec, by then ten times larger than Valley Alarm.

In the same month, SECOM entered into an agreement with Westinghouse to provide joint security services for American homes using community antenna television (CATV) networks. There were 1.7 million subscribers to Westinghouse's CATV service. The focus had been on leisure programming but Westinghouse wished to expand into the information field and was also searching for a partner capable of providing a high quality security service. The trust this would generate among subscribers possibly could create a foundation for growth. Iida had been studying the feasibility of using CATV networks to provide security services for sometime. The partnership was a natural one. From the knowledge gained in this venture Iida was able, the following year, to undertake the provision of domestic security services using CATV networks in three Japanese cities: Sendai, Mito, and Okayama.

With the takeover of Westec, Iida modified Nihon Keibi Hosho's electronic security alarm systems to meet American needs and production began at Westec. Response to the new product was good. Westec was now marketing a security package that integrated people and technology. It was a package that customers wanted at a time when specialization was the norm and manufacturing, monitoring, and guard dispatch were handled by separate establishments. With the new package, contracts, particularly for residential properties, increased markedly.

In March 1984 one particular event brought the Westec name to the fore. Actress Penny Marshall, returning to her Hollywood Hills home after a brief trip to the local store, spotted someone in the den. Thinking it might be her daughter she called out a greeting. When there was no response she sensed something was wrong and tripped the silent panic button on the burglar alarm. Because she had been absent from the house only a matter of minutes she had not bothered to set the alarm before she left.

As soon as the control centre received the signal from Marshall's home, two patrol cars were dispatched and were on the scene in less than seven minutes. A few minutes later they were joined by a Los Angeles police helicopter. In the meantime two 'rich kids' from Beverly Hills dressed up as Ninja warriors and brandishing a Samurai sword, confronted Marshall. Threatening to kill her if she did not cooperate, they ordered her to hand over 400 dollars in cash and sign an unknown amount of travellers cheques.

As they surrounded the house, the Westec patrolmen could see the teenagers scouring Marshall's home for valuables. Realizing they were about to be apprehended the teenagers released Marshall and fled through the back door into the hills and bushes. With the help of police dogs and the helicopter they were captured a short time later.[10] Penny Marshall played 'Laverne' in television's long-running 'Laverne & Shirley' comedy show. Because the show was being broadcast at that time Westec received nationwide publicity. The incident was covered by every American wire service.

Despite the efforts of the American management teams of both Westec and Valley Alarm, neither company was doing as well as Iida desired. He felt that the corporate philosophy had not been internalized. The turning point came with the installation of Michael Kaye as president and CEO of Westec in 1985.

Born in New Jersey in 1954, Kaye's interest in Japan was triggered when, at the age of fourteen, he spent a few days in Japan en route to Hong Kong to visit his uncle, Ira Kaye, who was engaged in the manufacture and sale of apparel. Kaye was fascinated by Japan. Shortly after entering Stanford University to study history he met someone who was studying the Japanese language and highly recommended the course. Kaye enrolled and found it very much to his liking. The following summer he participated in a two-month student exchange at Keio University in Japan. After a total of three years of Japanese language study at Stanford he applied to the Inter-University Centre for Japanese Language Studies in Tokyo in which Stanford University was one of twelve participants. The majority of Japanese scholars in America have studied at this institution at some point in their careers.

During the two years Kaye was at the Centre he 'passed the point of no return' in terms of his involvement with Japan. 'It was not a conscious, thoroughly thought out process', says Kaye. After returning to Stanford and completing his degree Kaye went to Harvard Law School. Upon graduation in 1980 he joined Gibson, Dunn, and Crutcher in Los Angeles, one of America's largest law firms. A prestigious firm, it had among its clients many Japanese corporations and was seeking to expand in this area. Kaye's fluency in Japanese and his understanding of the culture made him the logical choice to represent these clients.

During his time in Tokyo Kaye had become acquainted with

Yasuharu Nagashima, the main partner in the Nagashima & Ohno law firm in Tokyo. Nagashima had attended high school with Iida's brother Tamotsu and had undertaken some legal work for Iida while the latter was still in his father's employ. Later he had become the attorney for Nihon Keibi Hosho. Nagashima had taught at Harvard. While in Tokyo, Kaye had worked for Nagashima.

Shortly before Kaye graduated from law school Nagashima telephoned him to ask him which firm he was going to join because he had a client he wanted to send to him. That client was Nihon Keibi Hosho. Kaye subsequently represented the company in both the Valley Alarm and the Westec acquisitions. Kaye's tenacity and skill as a negotiator left a strong impression on Iida; at the same time the lawyer's respect for the Japanese entrepreneur was growing.

After three years with Gibson, Dunn, and Crutcher Kaye took a leave ostensibly to study law at Tokyo University. In fact, although he knew he wanted to get back to Tokyo, he was unsure of what he wanted to do.

> I remember being in the study at the law firm late one night. I looked around at some of the other associates working and I thought, 'It looks like these people are enjoying themselves more than I am.' I decided I needed to get out of the law field because to work that hard you really need to be enjoying yourself. Life is too short not to do something you enjoy. You spend more of your waking hours at work than any place else – with your family or doing anything else.

Kaye did enrol at the Law School at Tokyo University and at the same time worked for several clients including Nihon Keibi Hosho where he spent one day a week.

As his Japanese language fluency increased Kaye received more exposure to Iida's corporate philosophy and its implementation in everyday business dealings. As his understanding increased he began to question the nature of his role as a lawyer. To him it seemed that a lawyer simply utilizes specialized knowledge to accomplish a particular task within a confined procedure. So much of what they did, he thought, was to decide 'who gets what' either in a lawsuit or in negotiation. While Kaye acknowledges that lawyers have achieved much good in the United States and that law is an essential part of the American political fabric, he perceived much of their work to be nonproductive. In comparison the corporate manager's role appeared to be more creative.

After a year of study Kaye resigned from Gibson, Dunn, and

Crutcher and went to work for his uncle, Ira Kaye, a very successful American businessman in Hong Kong. His uncle wanted him to take over his business, Lark International, a diversified trading company. Kaye suggested instead setting up a branch in Tokyo. His uncle agreed. Kaye hired one Japanese employee, and the business was an immediate success. Kaye continued to spend one day a week at Nihon Keibi Hosho in an advisory capacity.

Towards the end of his second year in Japan Iida approached Kaye to seek his help to find a suitable American to install as president of SECOM's American operations. Such a person, Iida declared, should be intelligent with a good educational background, knowledgeable about Japan, and about thirty years of age. Iida's ear-to-ear grin made it obvious to Kaye that he was being offered the position. Kaye was flattered and honoured but his obligations to his uncle and the employee he had hired on the understanding of lifetime employment prevented him from accepting Iida's offer.

A few months later Iida approached Kaye again this time to ask him to find someone else suitable to head up SECOM's American operations. Now the qualifications were tougher. The person was to be in his forties, with considerable managerial experience. Kaye contacted two graduates of Harvard Law School who were among his acquaintances.

When each expressed considerable interest in the proposition Kaye began to re-evaluate his situation. He approached Iida and asked if he was still in the running as a candidate. Iida welcomed him with open arms. Kaye had only one stipulation – that the man who was working for him be allowed to come to SECOM with him. This person had given up a very good position with a large Japanese company in order to work for Kaye and Kaye wanted him to have the opportunity to return to a secure lifetime employment situation with a large company. Iida agreed and the arrangement has worked out well.

Ira Kaye was very upset with his nephew's decision and it was several years before the rift between the two men began to heal but Kaye is certain that the move was a good one. From a career perspective he knows that he did the right thing, that he is better suited to his new job than his old, even though, financially, he would likely be better off now had he remained with his uncle. In May of 1985, a few months after he joined SECOM, Kaye married a Japanese woman whom he had met while studying at the University of Tokyo. His uncle declined to attend the wedding.

Although Kaye was hired to be president and CEO of Westec and to spearhead SECOM's diversification in the United States, his first months were spent in Tokyo working for SECOM full-time, visiting all the different parts of the company and learning more about the security business. He needed to be well versed in the many facets of SECOM's operation if he was to be prepared to confront the challenge awaiting him at Westec. The company had been on the decline since the acquisition. Kaye's mandate was to turn it around.

In order to rebuild the failing corporation Kaye replaced most of Westec's senior management with stronger personnel. Iida feels he might have been less bold than Kaye in making such changes in part because Japanese business culture offers less flexibility in terminating employment, but he recognized that practices in the two countries differ, and he had given Kaye the authority to do whatever he felt was necessary to turn the company around. He had complete trust in the new president.

It is unusual in Japanese corporations to have someone as young as Kaye, and especially someone parachuted into the corporation, occupying as powerful a position as Kaye's. But his youthfulness has never been an obstacle at SECOM, Kaye feels, because of the way Iida set up the lines of command. The interface between Kaye and other SECOM employees is minimal as Kaye reports to Iida directly.

I'm sure that when I first came over here there were a lot of people who resented his decision to put me, an American and a young guy, in the position. And that's why I'm sure with some other Japanese companies, even if somebody wants to do that, they don't or they can't because of the consensus decision policy.

When you have a benevolent autocrat, if they are bright and capable that's great. If they are not you've got real problems. But when they're good, they're great because they don't have to spend time waiting for a consensus decision to evolve. Fortunately at SECOM we have a great person in that role.

Having represented a lot of Japanese companies in the States and knowing about Japan, with all due respect, I wouldn't want to work as an American at most American subsidiaries of Japanese companies. Typically, the pattern so far has been to have folks from Japan running the company either up front or behind the scenes with not a lot of authority in the hands of the local Americans. Japanese companies are not the only ones that do that. A lot of American companies in Japan do that but I really think Japan is particularly behind in that area.

Having said that, I understand this stance. Japanese do a very good job of communicating amongst themselves but for an outsider communication is difficult. It's not just the language, it's the culture and everything else. For me as the manager of SECOMERICA, if I were to go into Indonesia or Germany or anywhere else, I would need to have someone there reporting to me who I can talk to and who's going to understand what I want to achieve. The fact of the matter is there are very few Americans who can do that with Japanese.

So I understand the problem but on the other hand, as an American, I wouldn't want to work for most American subsidiaries of Japanese companies. It's just not very exciting, not very dynamic, because you're really outside the mainstream. You're a foreigner, you're a *gaijin*, you're really sitting on the outside.

I have some very close friends who are Japanese. When you are close friends differences fall away. But in a business setting many Japanese don't think of foreigners as real people. Not really *ichininmae* [fully developed people]. Foreigners are always kind of half people. The Japanese really don't think: 'This is someone who can understand me or knows what I'm about.' There's always a distance there, I think, generally. With Iida that was never there. It's not that he has lived in the States or speaks English. It's just something that seems to transcend that barrier that so many Japanese have.

SECOM is very different from most Japanese corporations. It's different because of Makoto Iida because he's very different and because he runs the company. It's not some kind of a fuzzy consensus. Because of the strength of my relationship with him I had a feeling that I would really have a free hand to do a lot here. With more than 3,200 employees and with two hundred million in revenues we have only three folks from SECOM and they're good people – young, energetic people in training positions out in the field – but not people in corporate management roles sent over to run the company.

Today, Iida seldom visits the United States more than twice a year, usually in January and in June – and then only for a matter of days – to keep the lines of communication open between himself and Kaye and to see for himself how business is going.

Under Kaye's direction new vitality surged through Westec. Even before the corporate mission statement was formalized through Kaye's visits to branches, talking with people, and making decisions, employees sensed the dramatic change in style, approach, and values.

Authority, for example, has been transferred down the line and placed as close as possible to the customers. Westec has become service driven. The customer is placed at the top of the organization chart with the people who service the customer next and everybody else after them. Employees have learned to think in terms of 'the customer' and to ask themselves what they are doing to service the customer.

Kaye has found that the best way to spread the word about the corporate mission is to talk about it to employees again and again. Says Kaye, 'It takes repetition, just like advertising.' With repetition the corporate mission began to sink in and employee behaviour changed. The company began to grow. When Nihon Keibi Hosho acquired it in 1982 Westec's contracts numbered less than 10,000. By 1990 these had increased to nearly 50,000. Among Westec's clients are many professionals in the entertainment field some of whom live in gated, access controlled communities. Westec secures the gateways and precincts of these communities as well as the individual residences within. As well, by 1990 the number of Westec employees had grown to 900 from 410 at the time of acquisition. Moreover, with the change in management employee motivation rose. Kaye comments:

> As in most businesses, the employees are members of a community which is a slice of real life. Society is made up of all kinds of people from those who are very bright and highly motivated to those who are not as bright or as highly motivated.

At Gibson, Dunn, and Crutcher, Kaye's associates were all, on a scale of one to ten, nines and tens. Kaye found them terrific to work with and enjoyed the 'esprit de corps' that existed within the group. At the same time, he found the situation unrealistic.

> In a community you have people who are at all different levels of ability. A company is more like this and I prefer it. I like the challenge of working with all types of people to help them reach their full potential – to be the best they can be.

Success in the American market led Iida to contemplate expansion not only in the security business but also into other peace of mind, service-related industries. To this end, in July 1987, SECOMERICA Inc. was established as a holding company to oversee SECOM's diversification in the United States. Kaye was appointed president and CEO. Diversification, says Kaye, 'was to be in service industries which delivered "peace of mind", Mr Iida's concept, to our customers.'

One of Kaye's first ideas for diversification was ambulance service. Feeling that this field came within the conceptual realm of SECOM's social responsibility, Iida was enthusiastic. As SECOMERICA's staff increased the personnel became available to examine seriously possible avenues of diversification. The ambulance industry offered much opportunity particularly in an aging society where the demand for medical transportation was likely to rise in the future. In April 1988 SECOMERICA acquired the emergency medical services division of HCA (Hospital Corporation of America), one of America's largest medical service companies. Several more acquisitions followed. With regard to the process through which corporate expansion is achieved Kaye says:

> It's fairly simple. No one is really going to tell you what to do. If you talk to investment bankers they're basically sales people; they have companies to sell and they want to sell you what they have. Some consultants can help you do some thinking but you really have to do it yourself.
>
> We looked at what we were trying to do – the 'peace of mind concept'. We developed our own mission statement here which is that SECOMERICA is a diversified services company whose mission is to deliver peace of mind to our customers through the delivery of innovative, essential-life services of extraordinary quality.
>
> The core concepts like 'peace of mind' around which SECOMERICA is built have come from Chairman Iida. We weren't interested in the hotel business or manufacturing. But within that framework, which was still very broad, my staff and I were able to design SECOMERICA's future.

Emergency medical services fit within SECOMERICA's mission. There were overlaps in areas such as marketing and operations. Looking at the ambulance industry as a whole there appeared to be a good opportunity for SECOMERICA to play a major role in the industry. Iida and Kaye established LifeFleet (at the time known as American Medical Transport), which today operates in Seattle, Tacoma and Spokane, Washington; Oklahoma City and Tulsa, Oklahoma; Pinellas and Hillsborough Counties, Florida; and in the Southern California area.

There are approximately 8,000 establishments providing ambulance services in the United States, including hospital ambulance services. Many of these are small-sized operations. Only 400 operate more than ten ambulances and only fourteen operate more than twenty. With 1,400 employees, LifeFleet is the largest provider of emergency medical services in the United States, averaging 1,000 transports per day in 300 ambulances. Total sales for 1990 were in excess of seventy million dollars.

Compared with Japan, regulations governing ambulance service in the United States are lenient. Ambulance service often incorporates emergency medical service provided by paramedics and emergency medical technicians. SECOMERICA's expansion into this area has provided SECOM with the opportunity to accumulate know-how which will be available for future application in Japan. Presently ambulance services in that country are tightly controlled by the government.

Initially under the auspices of the police department, since 1963 Japan's ambulance services have fallen within the jurisdiction of the national Fire Defence Agency of the Ministry of Home Affairs. The 4,440 ambulances that are kept ready round the clock at some 3,900 locations throughout the country are, first and foremost, a form of transport and their attendants are not equipped to provide medical aid despite growing public pressure. All medical services come under the Ministry of Health and Welfare. If ambulances are to be staffed by paramedics equipped to administer a degree of medical attention the regulation of the paramedics would come under the jurisdiction of this Ministry.

Doctors in Japan are divided between those who support the retention of medical treatment exclusively within their domain and see ambulances as transport vehicles only, and the majority, who favour the administration of emergency medical services by trained technicians, realizing that such services, administered at the hospital, often come too late to save the victim. In 1976 a computer controlled comprehensive emergency information centre was established in Tokyo by the municipal Fire Defence Agency to coordinate ambulance dispatch, available hospital beds, and available medical expertise.[11]

A 1989 survey conducted by the Fire Defence Agency of the Ministry of Home Affairs determined that in Tokyo the average elapsed time from the dispatch of an ambulance to its arrival at the hospital is 21.6 minutes. From its dispatch to its arrival at the accident site, however, the elapsed time averages only 5.8 minutes, leaving 15.8 minutes when trained staff could be administering medical aid.[12] Based on the findings of this survey, in August 1990 the Ministry of Health and Welfare's Emergency Medical Services Committee submitted an interim report to the Minister recommending that ambulance crew members be licensed to administer specified medical services.[13]

The state of ambulance service is far worse in rural areas than in the cities, with shortages the foremost problem. To fill the gap, in 1982 some taxi companies and undertakers began to earmark certain vehicles for ambulance service. More recently some 200 taxi corporations have begun to offer service to the handicapped, providing wheelchair and bed transport. The quality of both ambulance and handicapped service varies but entrants to the field continue to increase as public demand for expanded and improved service swells.[14]

In the United States, constantly rising medical costs, the development of technology which makes the in-home administration of many medical services possible, and the desire on the part of individuals for home treatment rather than hospitalization whenever feasible have combined to create a growing market for home health care services. In its search for areas for further diversification SECOMERICA examined health care services on a broader scope and identified home infusion therapy as the fastest growing segment.

Technological innovations have, over the past decade, enabled the number of illnesses commonly viewed as treatable with home infusion therapy to grow from less than fifty to over 1,000. At the same time, a growing number of physicians and patients are realizing that infusion therapy can be delivered at home where the familiar environment boosts morale and costs are 30 to 40 per cent less than in an institutional setting.[15]

There were a number of identifiable companies in this area the biggest of which, HMSS Inc., a rapidly growing, Houston-based provider of comprehensive infusion care nursing services across fourteen states, was publicly traded. Discussions with Kaye convinced Iida that the HMSS organizational culture would mesh with his own corporate philosophy. Iida was eager to see Kaye propel SECOMERICA into this new frontier. The new society he had visualized which would support this type of business had arrived, he felt, in the United States. After the customary critical business analysis of the attendant risks, Iida opted for the acquisition.

'Because the numbers were big', says Kaye, 'we did involve an investment banker to help us with the transaction. But the strategic thinking as to what to do – really nobody can decide that for you but yourself.' In November 1989, through SECOMERICA, SECOM concluded its largest acquisition to date: the $255 million purchase of HMSS. With revenues of $56 million in 1989 the company's annual growth rate for the previous four years had exceeded 50 per cent. This was the eighth largest Japanese acquisition of an American institution for that year. The largest, and most controversial among Americans, was Sony's takeover of Columbia Pictures.[16] Considerably less costly but causing no less consternation was the takeover of the RGI (Rockefeller Group Inc.) including the Rockefeller Centre by Mitsubishi Estate. Other acquisitions included CIT, Cook Cablevision, Lifomed, Shaklee, and AVX. SECOM was among the first Japanese companies to attempt an acquisition in the service sector.

Founded in 1982 HMSS operates through twenty branches, many with satellite facilities. About one-third of its 750 employees are registered nurses and pharmacists. The corporate goal is to improve the quality of life of the patient and the patient's family by delivering intravenous nursing and pharmacy services at sites other than hospitals, such as the patient's home. When it was founded, HMSS was one of the first health care organizations to provide chemotherapy administered in the home by a registered nurse clinician. In addition to chemotherapy, HMSS provides a number of other specialized infusion therapies, such as antibiotics including ampothericin B, cardiac drugs including dobutamine, growth hormone, pain management, and nutrition support for a range of patient specific needs in both the pediatric and adult population.

Specialists in intravenous therapy, HMSS's registered nurse clinicians also receive specific HMSS training to familiarize them with the company's standards, policies and procedures. Like HMSS pharmacists, reporting directly to the patient's physician they are 'on-call' round the clock seven days a week. In addition to on-site nursing service, HMSS personnel provide instruction for patients capable of administering their own medications. At the client's request HMSS will also take care of the paper-work involved in any medical insurance claims.

The company's network of branches and satellite centres makes possible the accommodation of patient referrals in many large markets and allows patients to travel if they so desire and still receive the same service. Sales for 1989 were US $56 million. Kaye's intention, with Iida's approval, is to expand HMSS qualitatively and quantitatively through a customer-oriented, efficient service that retains the human touch.

In October 1990, SECOMERICA acquired a majority interest in Care Visions Inc., adding a second company providing health care in a non-hospital setting. Care Visions specializes in providing comprehensive pediatric care to medically fragile children who need supplemental oxygen, use ventilators, or require constant treatment for neuromuscular diseases, heart diseases, feeding disorders, or cancer.

Care Visions began by providing home nursing care round the clock to these children so they would not require full-time hospitalization, which is both stressful and expensive for their families. In 1990, Care Visions opened their first Kangaroo Kids Center for Fragile Children where parents can have their medically dependent children cared for by nurses in a friendly, daycare-like setting during the day while they work. The name Kangaroo Kids was created because of the protective environment of a kangaroo's pouch for its young that are too vulnerable to survive on their own when first born.

The centres offer a socially nurturing environment for the children where they can interact with other children and perhaps feel less sick and isolated. It also provides families with an alternative to 24-hour nursing in their homes and greater opportunities to establish support networks with the parents of other fragile children. When the first centre opened, Care Visions also opened its own pharmacy to provide medications for its patients' intravenous medical needs. Since its association with SECOMERICA, Care Visions has begun an aggressive expansion programme to open additional home nursing offices and Centers for Fragile Children in the western states of the USA. Care Visions Inc. is now a wholly owned subsidiary of SECOMERICA.

Since its establishment SECOMERICA has grown from annual revenues of $40 million to its current $200 million. The company now oversees activities at sixty locations for its three divisions, security, emergency medical services, and alternative-site health care. From Iida and from experience Kaye has learned that service businesses are locally driven. The closer the locus of authority is to the customer in the field, the better the business will function. Accordingly, SECOMERICA's three divisions are managed fairly independently by their own presidents, although many opportunities are found to link the operations with group-wide efforts. The idea that authority should lie as close to the customer as possible is one that comes from the top and cascades down through the organization.

One of the reasons Iida enjoys American business, Kaye feels, 'is that you do have a lot more freedom to paint the canvas.' That kind of flexibility, to buy into a new industry, does not exist in Japan. Kaye continues:

> The nice thing about business here is that you can take a very large position in a new industry so you can get in immediately. The fun part comes after that – how you create that design. Not just how you take what you've bought but how you make it

better. That's what's fun. But you can start with a lot more here because you can go out and buy it.

However, just buying it is no good. Buying a company is easy. It's like becoming a father. But making a good company after you buy it is like raising children – it's much more difficult.

Kaye and his staff at SECOMERICA are still working to create the grand design. Their focus is on how to develop the fundamentals of the organization, its infrastructure, and its people. 'What I did at Westec', says Kaye, 'was really more what we call block and tackle – short- and medium-term things to improve the company and make it run better at what it always had been doing. Now we want to reposition – to go back and really do the design.'

This design, however, will not be made in isolation. Kaye is looking at improving all three lines of business within SECOMERICA simultaneously with the grand design. One project is the creation of SECOMERICA University with a core curriculum and programme that will orient people to the company and expose them to the broader picture of corporate-wide vision and values. At the same time it would facilitate employee development and aid in their formulation of a career path within the corporation.

As a first step Kaye hired a vice-president of corporate communications, Angela Sinickas, for in his opinion:

In service businesses the key is to have the vision clearly articulated; to have all employees understand that vision and how they fit into it; to make it clear to them what their job is and what they are supposed to do; and to give them as much authority as you can to do it.

SECOMERICA now has nearly sixty operation points – too many for Kaye to visit personally as frequently as he would like in order to assure internalization of the mission statement on the part of all employees. To supplement Kaye's face-to-face communication, staff are being hired whose sole responsibility is to communicate the corporate mission through the development of videos, newsletters, brochures, and other media.

What distinguishes a service business from a manufacturing business is the intangibility of the product. A tangible, manufactured product like a television can be tested before it leaves the factory. What will happen when the customer gets the television home is known in advance. When the product is a service like security there are hundreds of unsupervised contacts in the field that will determine customer satisfaction. These cannot be controlled by supervision. Rather the leader must convince the employees to become believers in the corporate mission and this belief must be reflected in their behaviour.

Kaye feels that he has 'a very long leash' within the SECOM organization. Iida has placed enormous trust in Kaye and his staff, a trust which, Kaye says,

> I take very seriously. It's very motivating. For the last year, because we've been particularly busy, we've been working seventy to eighty hours a week, week in, week out, Saturdays, Sundays. You don't get that kind of work out of people unless they feel very trusted and very empowered.

Kaye enjoys being able to build an organization that really is providing a new service. He likes to see people grow and considers it a privilege to be able to create an environment in which they can learn: 'I like the blank slate, the clean canvas that being at the helm of SECOMERICA offers.' SECOMERICA has diversified into many areas as yet unexplored by its Japanese parent, a reflection of Iida's recognition that, given the cultural and linguistic gaps, allowing Kaye nearly unfettered control is a realistic approach.

SECOM's involvement with new dimensions of the social system industry in North America will provide the necessary accumulation of know-how so that the company will be ready to offer this type of service in Japan when the now tightly controlled medical industry is deregulated in the near future. Japan's rapidly aging population is expected to offer a ready market for private home health care services. With SECOM's well-established monitoring and dispatch systems already in place expansion into the emergency and alternative-site health care fields seems feasible.

THAILAND

With the rising Japanese yen in the mid-eighties direct investment by Japanese corporations in NIES (newly industrialized economic states) and ASEAN (Association of Southeast Asian Nations) countries increased. Among the recipients of this investment was Thailand, one of the most politically stable Southeast Asian nations. Japanese investment in Thailand in 1986 totalled 1.7 billion baht, a three-fold increase over the previous year. Investment on the part of other nations was also rising. The industrialization of Thailand was taking off.

Iida and Sugimachi were well aware of the fact that the crime rate rises as industrialization advances. In Thailand measures to counteract burglary were not well developed, barred windows being the primary deterrent. Of questionable effectiveness, barred windows also hindered escape from fire. While receptiveness to a private professional security service was low, Iida felt the demand could be cultivated. SECOM's systematized business know-how accumulated through domestic and foreign operations was, Iida thought, sound enough to be implemented.

Iida and Sugimachi travelled to Thailand in early April 1987 and confirmed their belief that the country would prove to be a fertile area in which to expand the security business. Iida set his sights on a joint venture and, on his return to Japan, dispatched Yasuo Ezawa to Thailand to search for a suitable partner. Ezawa had worked for Komatsu Limited, the world's second largest manufacturer of construction machinery next to Caterpillar of the United States, for twenty years before joining SECOM in 1984 at the age of 49. As the person responsible for opening Komatsu's first overseas office – located in Belgium, and later posted to Lebanon – Ezawa had acquired considerable knowledge and skill in founding and running overseas operations.

Ezawa had no time to prepare before he left for Thailand in mid-April 1987. Among other things, until his departure, he had been occupied with preparations for and the reception of the Minister of the Interior of Saudi Arabia, brother to the king, who had chosen to visit SECOM while on a trip to Japan. Says Ezawa:

> This was very much SECOM's style, when a decision was made we would act immediately. I had to get ready to go to Thailand on the fly. Time was too short to obtain a visa so I could stay only two weeks.
>
> I still remember the day I arrived at Bangkok airport. It was over 40°C and humid. I was sweating so much I felt I was standing under hot water. I couldn't speak the language but I was ready to take up Mr Iida's challenge.

With the help of personnel from Japanese financial institutions in Bangkok, Ezawa identified several industrial conglomerates as possible partners. Most of these were owned by overseas Chinese. He collected data with regard to several important aspects including managerial strength, sales, marketing strength, and size and analysed it in order to narrow further the search. He visited half a dozen of these corporations.

Interest in security was very low as was the market permeation of insurance. Because labour costs were low, hiring guards was easy and there was no perceived need for machine security. The people of Thailand are conservative. Few people with whom Ezawa spoke could grasp what it was he was trying to suggest. Ezawa eventually identified three corporations suitable for a joint venture but did not hold out much hope for the acceptance of any of them. Ezawa recalls his distress:

> For the remainder of my stay I could not sleep at night knowing I had not completed my mission. I thought the Saha Pathana Group or either one of two other corporations would be a good partner but interest on the part of all three was slight. I could not meet with the chairman of Saha Pathana but I did spend time with a high ranking executive who had studied in Japan and he expressed a slight interest in the proposal.

After conferring with Ezawa on his return from Thailand, Iida concluded that the Saha Pathana Group would be the appropriate partner. The Saha Pathana Group consists of nearly 200 corporations with sales totalling 1,500 billion yen. Employees number 15,000 many of whom, prior to joining Saha Pathana, received scholarships to study in Japan. The company's activities include the production, distribution, and sale of a wide range of consumer products. Connections with Japan are strong through joint ventures with Lion, Japan's largest producer of dentifrice and synthetic detergents; Wacoal Corporation, Japan's leading manufacturer of women's underwear; and Janome Sewing Machines. Saha Pathana entered the high-tech field through a joint venture with Samsung of Korea for television assembly.

Shortly after Ezawa's return from Thailand he learned that the chairman of Saha Pathana, Boonsithi Chokwatana, would be visiting Japan mid-May. It was Boonsithi Chokwatana's grandfather, coming to Thailand from China, who had established a grocery store that later, under the direction of Chokwatana's father, developed into a large corporation. As a young man, Chokwatana had been sent to Japan to gain experience working in a Japanese company and to learn the language. Although his schedule was tight, Chokwatana made time to accompany Ezawa on a tour of SECOM's Research Institute and Control Centre and to meet with Iida.

The two men conversed for about one hour. Within the first ten minutes they had agreed to enter into a joint venture. Ezawa feels that Chokwatana made the decision to participate because he realized the future possibilities and challenges of such a venture. He knew that profits would not be produced immediately, that it would take much hard work to make a success of the new company.

Just as Iida chose an auspicious day to establish formally Nihon Keibi Hosho, he asked Chokwatana to choose a special, lucky day to launch the joint venture. September 9 was the date selected. This allowed only three months for the necessary preparations to be completed.

Iida appointed Ezawa director of the five-member project team created to launch Thai SECOM Pitaki. The team made several trips to Bangkok to study the area. They learned that private lines were in high demand and short supply and that there was a similar imbalance in the supply of and demand for public telephone lines. Roads were congested and the team had to consider the mode of transportation most appropriate to particular areas.

Bangkok is situated on a flood plain. During the rainy season communication lines may be out of commission and work often grinds to a halt. This necessitated the development of countermeasures to enable security systems to operate under such conditions. The project team also had to determine the ideal location for the control centre.

Trainees from Saha Pathana were sent to Japan to become familiarized with SECOM's systems and operations. Various manuals had to be adapted to suit the Thai situation and translated into the Thai language.

As this was the first security firm to be established in Thailand there was no precedent for the granting of government approval. At first it appeared that this would be a stumbling block but eventually the Ministry of Commerce issued the licence necessary to establish the joint venture. The joint venture contract was drawn up and signed by both companies and SECOM supplied 49 per cent of the investment capital. Thus, the company was launched. But there were more hurdles ahead.

The president and most of Thai SECOM's staff were to be Thai with Japanese personnel providing the know-how, at least in the initial stages. In addition to Ezawa as vice-president and general coordinator, the intention was to send Japanese personnel who could assist with the installation of the communications infrastructure, marketing and sales, and 'office management'. Work permits were required by all of these individuals but the Thai government would only issue two in addition to Ezawa's. Had the joint venture been in the manufacturing sector and geared to export it would have received preferential treatment but being in the service sector this was not the case.

Eventually, after four months and thanks to the intervention of influential 'friends' this problem was overcome and the appropriate personnel were sent from Japan to Thailand.

All hardware had to be imported from Japan. Because nothing similar existed in Thailand appropriate classification categories had not been established by customs officials for assessing import duties. SECOM officials faced the challenge of explaining the purpose of the equipment to customs officers and preparing satisfactory documentation. Next, permission to operate the security system had to be obtained from the Telephone Organization of Thailand. This done, Thai SECOM was ready to hire employees.

The company was competing for personnel against numerous foreign firms that had recently established branches in Thailand. At the time of establishment there were some 500 Japanese firms registered with the Japanese Chamber of Commerce in Bangkok. Thai SECOM's name and product were not known and its future was uncertain, making recruitment more difficult.

New employee training, particularly of sales personnel was, and continues to be difficult. Ezawa explains:

> Sales employees receive instruction in such matters as human interaction, sales planning, salesmanship, cost estimates, and contract negotiation. Often the instruction has to be repeated more than once before concepts are grasped. Once on the job, employees often quit within two or three months.
>
> The climate of Thailand can be described as hot, hotter, and hottest. I think the heat drains a salesman's energy. The tempo is very slow, nothing like the fast pace we're used to in Japan. Inattention during the training sessions contributes to the inability to successfully explain the product to potential customers. The subsequent rejection of SECOM's service by these potential customers leads to a lack of self-confidence on the part of sales people.
>
> We try to explain that if employees stay with the corporation and work very hard they will move up the ladder just as their counterparts in Japan, Taiwan and Korea have done. After such a 'pep-talk' salesmen will often increase the number of corporations they visit in a day but, before long, will quit.
>
> I remember an instance in the early period when this was exactly what happened with eight salesmen on one day. They simply failed to show up for work and never came back. Advertising positions, interviewing, selecting, training, assimilating into

the workforce, then looking for replacement personnel – the cycle goes on.

Retaining guards has proved challenging as well. At the outset there was widespread aversion to wearing uniforms and particular dislike of corporate badges. It took time for the guards to appreciate the function of the corporate symbols on the uniforms and patrol cars.

Ezawa learned that the Japanese way of doing business could not be imposed without adaptation to the Thai cultural environment. What was common sense in Japan was not necessarily common sense in Thailand. Thais are very reserved. They mask anger and sadness with a smile. They are a kind and gentle people. Public criticism is uncommon. Reprimands are given one to one in a kind manner. Says Ezawa:

> There is a story circulating among Japanese executives in Thailand about one of their number who suggested to his Thai secretary that they should buy some goldfish to decorate the office. The secretary replied that the loudness of her boss's voice would surely kill the fish.
>
> After a while I realized that the Japanese at Thai SECOM were like mice on an exercise wheel. They were working very hard but going nowhere. The Thais, on the other hand, were like a slow moving river. When we tried to impose the Japanese way of doing things we would meet with resistance and work would not proceed.

Ezawa learned to speak the language in order to develop heart-to-heart communication with his co-workers. In his private life he sought out the positive and enjoyable aspects of Thai culture. With Ezawa's guidance Thai SECOM is gradually taking hold. With fifty employees, one control centre, and six depots, by the spring of 1990 the company had 100 contracts, sixty of which were with Japanese corporations. The remainder include several corporations belonging to the Saha Pathana Group. Contracts with unrelated companies have been increasing since October 1989, however. With the Thai economy booming guard salaries have been on the rise making machine security like that offered by Thai SECOM increasingly advantageous.

With the groundwork completed and the economic climate favouring the company, Ezawa has returned to Japan and Shinjiro Tsushima has replaced him as vice-president of Thai SECOM. Says Ezawa:

> In the spring of 1987, after returning from Thailand, Mr Iida was certain that the future of the security business in Thailand was

bright. To be honest, I was unsure. Now I know that he was right. The market in Thailand is very bright. I admire Mr Iida's foresight and his unwavering belief in his ability to succeed.

Early each year the most senior Japanese executive in each of SECOM's Southeast Asian operations meets with the others and with Sugimachi, director of SECOM's international operations, to share information, report progress, and receive Iida's future objectives for the international scene within the context of SECOM's total operation.

Through overseas ventures SECOM has come to understand the necessity of more comprehensive language and cultural training for personnel to be sent abroad. There is an awareness also of the need for more positive, company-wide cooperation and respect and appreciation for local cultural patterns and business styles.

11 Research – foundation for growth

RESEARCH CONSOLIDATION

SECOM's successful expansion into many innovative fields stems from Iida's realization that innovative endeavours are not merely a 'function'; they must be organized as a 'business'. Research, product development, manufacturing, and marketing are perceived not as existing in their traditionally conceived sequential order but rather as integral parts of the holistic process of new business development, each to be embarked upon at a time appropriate to the specific situation rather than in accordance with a preconceived time frame. Iida thought that the establishment of a substantive research and development foundation was essential in order to lay the groundwork for new business opportunities.

Until the late 1970s research and development was a section within the company's main office. Its personnel numbered about a dozen. Akira Satori, today manager of the systems design department of SECOM's Technical Centre, came to the company from Oki Electric and played a significant role in the evolution of the research and development department. The SP Alarm System developed by Shiba Electric utilized private circuit lines and worked well within the Tokyo metropolitan area. The system needed improvement, however, if it was to function elsewhere. Iida consulted with NTT personnel and Oki Electric was identified as the company most likely to be capable of meeting his requirements. Soon after, Satori joined Nihon Keibi Hosho. Says Satori:

> SECOM's research and development strategy has been conceptualized almost entirely by Mr Iida. As long ago as the early seventies he would talk about the future liberalization of telecommunications and his plan for production facilities integrated into the total corporate scheme. Everything he told us has been actualized.

In order to accommodate research aimed at expanding the company's frontier, Iida established the Engineering and Development Centre in 1979. It was housed in a building purchased some years before in Musashisakai on the outskirts of Tokyo. It was here that SECOM's small but enthusiastic research team together with such advisers as Tokujun Totani, an expert in security hardware with the Tokyo Police Agency, held a series of seminars. These seminars were the initial step in corporate technological expansion.

As Iida's conception of his business expanded into the information technology field he saw the need for a larger, more inclusive research facility. Centralization of the research laboratory under head office was the dominant practice. Such corporations as Hitachi, Matsushita, Toray, and Honda all followed this trend.[1] The ensuing search for an appropriate site led to the purchase of a Sansui Electric factory located in Mitaka on the outskirts of Tokyo. Sansui Electric is a medium-sized audio maker excelling in amplifiers. When the home audio market slumped and Sansui also ran into trouble with its car audio equipment it sold the Mitaka factory in an attempt to streamline production.

At the time Nihon Keibi Hosho purchased it the factory premises were rundown. Nihon Keibi Hosho renovated a portion of the building and established a Technical Centre there in 1981. This was a temporary arrangement until a new centre was opened in 1986. The Technical Centre absorbed the Engineering and Development Centre. Research activities that had been dispersed across the country were now gathered together under one roof permitting tighter coordination and the establishment of a research data bank. Centralization created an environment which fostered the development of new products not related to the current production. It made it possible for the laboratory to develop a long-term view not subject to profit pressure. The Centre's mission was threefold: to develop new products; to decrease the number of false alarms in the field; and to secure the dependability of equipment provided by Nihon Keibi Hosho.

The establishment of the Technical Centre coincided with Iida's request that research begin on the wireless transmission of sensor signals. False alarms still occurred occasionally with the SP Alarm System resulting from problems with the wiring linkage between sensors and the on-site controller. To rectify the problem Iida chose not to improve the weak link in the system but rather to eliminate it; an innovative approach. The solution developed by the Technical Centre was to convert the signal transmission medium from wire to microwave using lithium batteries as the energy source. Above a certain strength, microwave transmission was subject to strict govern-

9 SECOM's Technical Centre, Tokyo, completed in 1986.

ment regulation. The challenge confronting the Technical Centre was to build a system dependent upon microwave transmission of a frequency weaker than that to which government regulations applied in order to avoid interference with or from other microwave usage. This innovative solution to the problem had the added advantage of eliminating the cost involved with wire installation.

All sensor-related research formerly conducted by various sections could now take place at the Technical Centre. A variety of sensors were developed. Some failed in the experimental stage, others appeared to work well in the testing stage but failed in the market place. Still others were relatively successful but required ongoing improvement. Tackling increasingly complex problems necessitated more sophisticated facilities, one example being a microwave shield darkroom. In 1986 an entirely new building was opened on the Mitaka site. The 8,800 square metre complex has four storeys above ground and two basement levels and today houses SECOM's IS (Intelligent Systems) Laboratory as well as the Technical Centre.

Iida looked to NTT to find an appropriate individual to head the new Technical Centre. NTT's research institute, Denki Tsushin Kenkyujo, commonly known as Tsuken, is the nation's leading information communication and electronics research institute. Employing over 6,000 researchers, Tsuken attracts the country's top ranking graduates from leading universities. Recently the institute has spent in the neighbourhood of 240 billion yen annually on research and development. Personnel, information, and research funds are abundantly transferred to the private sector to the extent that problems sometimes arise from the excessively close links between NTT and private industry.[2] Upon retirement from NTT it is not uncommon for employees to find positions in the private sector. As well, in order to provide opportunity for advancement for younger workers older workers are encouraged to seek other employment. NTT's scientists provide a constant source of highly qualified researchers for private industry.

From among them Iida recruited Saiji Miyauchi. Graduating from Tokyo Metropolitan University in 1960 with an advanced degree in electrical engineering, Miyauchi joined Nippon Denshin Denwa Kosha. During his twenty-three years with the government organization he was involved in telephone circuit and telephone exchange design, research into the development of mega scale data communication systems, and the investigation of future telecommunication demands. His technical and administrative skills were well honed but he felt that his innovative and creative talents had not been allowed to blossom in NTT's rigid atmosphere. SECOM, on the other hand, would nurture these gifts by allowing him free reign in the development of the Technical Centre. He decided to make the move. Iida, Miyauchi, and SECOM's executives summarized the purpose of corporate research and development in the following guidelines:

1 To develop innovative unique systems.
2 To develop high value added systems.
3 To develop marketable systems in a timely manner to meet societal needs.
4 To develop systems with high dependability.
5 To develop user friendly systems with appropriate human-machine interface.
6 To develop systems which are easy to install, maintain, and repair.

From the outset Iida has perceived innovation as a distinct and major task. For him, innovation is an economic and social term rather than a technical term. It is measured in terms of its effect on the environment and is therefore always market focused. Guidelines three through six reflect this view.

Miyauchi has organized the Technical Centre into four departments: Technical Planning, Product Development I, Product Development II, and Quality Control. The Technical Planning Department consists of two sections: Systems Planning and Technology Management. The Systems Planning section incorporates activities such as the conceptualization of prospective systems as merchandise to be marketed; the preparation of basic product/system specifications; and the coordination of research and related activities towards the production of marketable merchandise. The Technology Management section is responsible for the formulation and/or compilation of technology standards; for all patent and consignment contract related activities; for technology related liaison with government offices; and for the development of manuals to accompany new products or systems.

Product Development I consists of three sections. The first is concerned with the development of institutional centralized security systems by type and size of institution. The second section concentrates on the development of security systems for households, apartments, and condominiums, on the development of medical alert systems, the development of HA/HE telecontrol and the advancement of cable television technology. Section three is concerned with the development of terminals attached to alarm systems and the development of local alarm safety equipment.

Product Development II is also divided into three sections. The high security project section deals with the development of high technology-based security systems for specialized usage. SECOM's primary focus has been commercial and domestic security. Iida was aware of the demand for improved security of national defense and of the nation's critical infrastructure. SECOM has begun to move into this area. Users include atomic energy plants, airports, military bases, hydroelectric dams, national borders, fuel storage areas, jails, etc. Among the numerous projects undertaken are the development of night-time surveillance systems, electronic fences, and the development of totally integrated building management systems.

The technology acquired has a spin-off in private sector security. The sensor section is responsible for the development of all sensors. These include heat sensors, ultra red ray sensors and smoke sensors for fire detection all of which have versions for internal and external use; indoor gas sensors for gas leaks and incomplete combustion; magnetic and limit sensors to detect the unauthorized opening of doors, windows, and shutters; ultra red ray sensors, image sensors, electromagnetic wave sensors, supersonic wave sensors, and pressure sensors to detect approaching human beings; and glass sensors and vibration sensors for the detection of the destruction of walls and safes. This sensor technology has application in building facilities management and in medical diagnosis and patient management, areas into which SECOM is expanding. The focus of the third section of Product Development II is the development of on-site fire extinguishing equipment including sprinkler systems and substitutes for Halon gas extinguishers in both automatic and manual versions.

Since the discovery of the destruction of the ozone layer in the mid-eighties and the identification of chloroflurocarbons (CFCs) as the major ozone-destroying chemicals, the widespread usage of Fluorine and Halon gases has become an issue. Because they are highly effective and create little secondary damage, Fluorine and Halon gases have been used extensively in fire extinguishing equipment at museums, libraries, computer rooms, and other sites where the damage rendered by water could be as devastating as that resulting from fire. In 1987 twenty-four nations met in Montreal and agreed to cut the use of CFCs by half by 1998. Japan has now ratified the agreement and SECOM, like other Fluorine and Halon gas users, is looking for suitable alternatives.

In the security service industry the quality of the hardware is crucial. Unlike equipment that is in constant operation, security hardware is seldom required to operate. When it is activated, however, it must be in top working order. A very specialized kind of dependability is demanded of security hardware. Iida saw the need for facilities that would permit tight quality control. The Quality Control Department was established, consisting of two sections: the technology dependability section and the inspection section. The former deals with the evaluation of the dependability of developed security systems, the development of product quality dependability criteria, and the development of measures to counteract the problems which have resulted in the rejection of particular security systems. The latter section deals with product quality inspection and the compilation and management of inspection results.

In this section all prototypes for off-shore manufacture are tested. These include those developed in Japan and those developed abroad to suit particular local conditions. In Japan, SECOM tests all components before installation. Abroad, however, quality checks on finished products are not as thorough and are in the process of improvement. Today products produced off-shore are used in the country of manufacture only because of differences in quality, cost, usage conditions, government standards, tax, and maintenance measures.

In addition to the four departments that make up the Technical Centre there are two offices and a special project unit known as the ES (Emergency System) Project. The Office of Technology Investigation identifies new areas for technological research and development, cultivates concrete applications for this technology, and analyses and evaluates its commercial potential. The second office is the office of technology application. It investigates the application of advanced technology and evaluates core technology and new materials as well as conducting basic research into future security communication systems.

The objective of the ES project is the development of information communication via electromagnetic waves. One of the products SECOM has been marketing since November 1982 is 'My Doctor', an emergency alert system to be manually activated by its wearer when a problem arises. This system works well so long as the individual remains on the premises which are linked to the home controller and thus to SECOM's control centre. When the alarm is activated off the premises, however, the signal may not be picked up by the control centre and, if it is, locating the position of the wearer is nearly impossible.

The ES Project is devoted to bringing about a marriage of electromagnetic waves and tracing technology and the development of a commercially viable alert system that would allow the user free range of movement. Technically this has been accomplished and the need for the product exists. The cost of manufacture, however, along with regulations governing the use of electromagnetic waves have so far proven to be stumbling blocks to market introduction. Once these obstacles have been overcome the product will have application far beyond 'My Doctor'.

In all, some two hundred SECOM employees are involved in the activities of the Technical Centre working toward the development of a comprehensive social system industry with security at its core. Says Iida:

> SECOM's security systems must be defect free. Consider the sensor. Regardless of the level of sophistication of the computer-based man/machine security system, sensor imperfections will destroy its dependability.

The sensor gathers and transmits information, functioning similarly to the five senses of the human body. It must be dependable, long-lasting, and allow for easy checks to ensure that it is functioning. It must incorporate an automatic malfunction detection mechanism. The structural design must provide for easy installation and minimum maintenance and the sensor itself must be unobtrusive in appearance and blend in with the surroundings. As well, it must be economical, energy efficient, and in continuous supply. With recent developments in the microelectronic field sensors are becoming increasingly sophisticated.

Presently SECOM has some 270,000 domestic contracts encompassing twenty-five million sensors. If each of these were to malfunction even once a year incorrect information would be transmitted to the control centre approximately 68,000 times a day. Even if they malfunctioned only once in ten years inaccurate information would still be received 4,100 times a day. Such malfunctioning would require the unnecessary presence of a beat engineer to investigate a non-existent problem. Plainly this would play havoc with the efficient operation of the company. Far worse would be the failure of a sensor to respond in an emergency and for the problem to go undetected until it was too late.

False alarms are a problem world-wide. The city of Calgary is one of the few urban centres in Canada to have an alarm byelaw. Anyone installing any kind of alarm system in a home or business must first obtain a free city permit. There were 4,296 alarms registered in Calgary as of April 1990. In 1989 there were more than 2,000 false alarms in the city. When three false alarms are registered at one location in a calendar year, the owner receives notification in the mail to correct the problem. After five false alarms a police constable is sent to the premises. After the seventh false alarm the permit is revoked and the owner ordered to shut down the system.[3] The idea of an alarm malfunctioning seven times in one year is, for Iida, unthinkable and he is striving tirelessly to reduce the failure rate.

Constant improvement in SECOM's sensors and all other equipment stems from the Technical Centre's own initiative and also from information obtained from the market through the sales department, beat engineers, and the Control Centre where customer inquiries and complaints are directed. The results of systematic computer analysis of maintenance and repair records are also integrated into product improvement and development.

LONG-RANGE RESEARCH

The establishment of the IS (Intelligent Systems) Laboratory was aligned with SECOM's future product scope. Initially, all basic research and technology application development was carried out by the Technical Centre, however, since the opening of the IS Laboratory in December 1986, long-term research projects, that is, commercial applications which are more than three years into the future, are handled by the IS Laboratory.

The IS Laboratory was established within the same building as the Technical Centre for the purpose of researching intelligent information processing technology necessary for what Iida conceptualized as the 'social system industry'. He felt that the technology the company had acquired in the past with regard to the security business was insufficient to propel the corporation forward. As the security system became more sophisticated and its use expanded, the control of equipment by artificial intelligence, the introduction of sensors activated by image and voice processing, and the utilization of a letter and pattern image voice integrated network was necessary. The newly created laboratory would be in the forefront of developing the system's infrastructure.

To establish the laboratory Iida sought a director competent in both the technical area and in administration. Again he looked to NTT, this time selecting Shinichiro Hashimoto. Hashimoto graduated from Keio University in 1958 with a degree in electronics engineering. He worked for NTT for twenty-eight years until joining SECOM early in 1986. His primary concern at NTT had been fundamental research into speech and image processing. In 1972 he had received his doctorate from Keio University after writing a dissertation on voice imitation in multi-frequency receiver signal systems. He was also involved in administration. At the time he left NTT he was the head of the Visual Communications Department of the Electrical Communications Laboratory in Yokosuka.

Iida discussed with Hashimoto the purpose behind the establishment of the research laboratory but gave him no firm instructions as to what form it should take. Hashimoto's aim was to create a research laboratory that would not only accommodate SECOM's needs but would compare favourably to research institutes throughout the country. Hashimoto drew up a master plan. Included in it were the purpose, goals, and operational guidelines of the laboratory, the role of the laboratory with SECOM's overall corporate strategy, the internal organization structure of the laboratory, the personnel system including salary scale and promotion criteria, the policy regarding the external relationships of the laboratory, the reporting and communication systems, and the education and training programme for laboratory personnel.

Hashimoto presented his proposal to Iida. 'Iida made no detailed comment. I felt he was assessing my behaviour. He then turned to

Miyauchi, the Director of the Technical Centre and said, "I think the plan will succeed." The proposal was scrutinized by the Laboratory Preparation Committee of which Hashimoto and Miyauchi were two of the six members. Iida gave final approval to the plan in July 1986. From among the suggestions submitted by Hashimoto, Iida chose the name Intelligent Systems Laboratory and it was launched officially on 1 December 1986.

The laboratory was to be housed within the Technical Centre and sufficient funds were to be made available so that Hashimoto could follow through with his plan. The major problem to be overcome was the lack of human resources. At the outset Hashimoto was assisted by three researchers, one who had come with him from NTT and two from within SECOM. No one else from within the company could be spared from their jobs to move to the laboratory so Hashimoto had to look elsewhere for recruits.

The competition was stiff. Throughout the eighties the average annual increase in research expenditures in Japan was 8.2 per cent, putting the country neck and neck with West Germany as the pacesetter in research and development. As R&D budgets grew so did the number of research personnel, increasing 5.2 per cent annually over the latter half of the decade. Six out of every 1,000 Japanese workers are now employed in research and the country has the largest number of researchers per capita of any in the world. In 1987, 76 per cent of all research money in Japan was spent by industry with universities and government organizations spending 13 and 10 per cent respectively.[4]

In 1987, as SECOM had not yet established a research tradition, it was difficult to attract personnel. Hashimoto knew that the success of the laboratory would depend upon the calibre of the scientists he employed. He focused his recruitment efforts on undergraduate and graduate students at major universities throughout Japan.

During his years of employment with NTT Hashimoto had been a regular participant in academic conferences enabling him to develop ties with professors and graduate students in the sciences. Wishing to strengthen and expand these ties Hashimoto visited a number of universities discussing the Intelligent Systems Laboratory at length first with faculty members and then with interested students. This has become his standard pattern of recruitment.

Hashimoto devotes April through June each year to visiting about twenty major universities, meeting with some 160 faculty members and a similar number of students recommended by their professors. In the course of these visits he also lectures and participates in seminars.

During the summer these students visit SECOM. Hashimoto spends three hours with each student showing them the research facilities and describing the academically challenging work and favourable working conditions. Flexible hours accommodate working at home if desired, the salary is higher than for most other comparable research jobs, funding for research expenses is generous, the facilities are modern and well equipped, and participation in academic conferences both at home and abroad is encouraged.

Hashimoto then takes the visiting students to NTT's research institute and encourages them to visit other research institutes on their own so they can objectively evaluate the Intelligent Systems Laboratory. If, after all of this, they feel that SECOM is where they would like to work he invites them to start work the following 1 April, the traditional starting date for all new employees in Japan.

Since the Laboratory was established in 1986 there have been four opportunities to hire researchers. Each year about fifteen competent researchers have been recruited by SECOM bringing the number of researchers to over sixty. Half of these hold graduate degrees. They have been attracted to the corporation by such factors as SECOM's contribution to society through the provision of security measures; its solid research foundation and support of individual research projects; the opportunity to continue with research comparable to that carried out in graduate level academic institutions; and the realization that they could be instrumental in shaping the future of the IS Laboratory.

Hashimoto's intention is to recruit and retain top-notch researchers. At the same time he must guard against the development of an elitist group, a realistic concern in research department management. It has been pointed out that the research department may be exceptionally hard to manage. There is a tendency for it to develop a culture 'deviating from a company's norms'.[5] To immerse researchers in the corporate culture, Hashimoto, from the outset, provides a series of training sessions in institutional settings.

New recruits in the IS Laboratory are given a week-long orientation session composed of lectures and tours to introduce them to all facets of SECOM's operation. This is followed by two weeks spent visiting locations where SECOM's various security systems are in use. After this comes two months in the Technical Centre devoted to familiarizing the new employees with technological developments related to SECOM's products. At the end of this time the employees are assigned to their specific research areas. From time to time researchers are given further opportunity to mix with employees from other divisions.

Researchers in the laboratory will number two hundred by 1991, including some non-Japanese. Hashimoto set the number of foreigners at 5 per cent but Iida feels it should be closer to 20 per cent. To date twelve foreigners have been employed at SECOM's Research Centre. Of these, four are still there. Living in an island country in close-knit communities Japanese tend to be alike in their thinking and behaviour. Bringing in foreigners exposes them to new realities. Japan is presently grappling with the problem of internationalization. To have foreign researchers working beside young Japanese scientists promotes cultural understanding and breeds a certain 'familiarity' and comfortableness with outsiders that is difficult to develop without ongoing contact. Being at ease with foreigners is essential, Iida believes, if SECOM's personnel are to mix and integrate in the international arena. There is much to be learned in the area of software technology from countries like the United States and the presence of foreign researchers would benefit SECOM in this area as well, Hashimoto points out:

Hashimoto recruits foreigners from a number of sources. Since 1986 the American Electronics Association has sent ten Americans annually to Japanese corporations. After a two-month immersion course in the Japanese language the participants are sent to work in major Japanese corporations such as Hitachi and Toshiba. Hashimoto has applied every year and to date SECOM has received three masters level science graduates. Each has stayed for a year. Janelle Jurek, a graduate of the University of Minnesota with a Masters degree in information engineering was with the company in 1988/89. Her interest in Japan was sparked by a course she took in Oriental Civilization. The instructor spoke of the excellence of Japanese software technology and triggered an interest in science and technology in Japan. She passed the Japanese Studies Fellowship Exam and was chosen to spend a year as a researcher at SECOM's Laboratory.

The Massachusetts Institute of Technology (MIT) also sends annually ten graduates to Japan. Since 1988 SECOM has received one of these researchers each year for a twelve-month stay. In 1989 Kyo Pochen joined the Laboratory for a year. He applied to the MIT Japan programme because he wanted to learn more, not only about Japanese technology, but also about Japanese culture, history, and the way Japanese think. MIT assigned him to SECOM. He was well satisfied with his posting as it gave him the opportunity to continue to study image information processing and also to live in Tokyo.

Hashimoto also tries to recruit foreigners through the network of associates he has developed through attendance at academic conferences. While the foreign scientists by and large enjoy their work at the

Intelligent Systems Laboratory they face considerable difficulty with communication even though SECOM provides language training for them at an external language school. As the number of foreign researchers increases SECOM is contemplating offering in-house, ongoing Japanese language classes more quickly to breakdown communication barriers.

Hashimoto has organized the Intelligent Systems Laboratory into four research divisions and a planning and administration department. The research divisions are artificial intelligence, voice and pattern recognition, image and signal processing, and fibre optic communication systems and robotics.

Research themes are determined by individual researchers in consultation with supervisors. The themes must be in accordance with the objectives of the Intelligent Systems Laboratory. Researchers are involved in a diverse variety of research. In the artificial intelligence section for example, research is being conducted into the development of a system that will permit data retrieval through the use of 'natural' language. Janelle Jurek was involved in the construction of a three-dimensional spatial information model which would allow for the optimum placement of sensory instruments and other equipment within the structure under surveillance.

In the speech processing section work is underway on speaker identification (machine identification of an individual by voice analysis); voice synthesis to facilitate the accurate mimicking of a human voice by machine; and voice scrambling. With the introduction of the cordless telephone and the car phone, portable communication systems have gained widespread popularity. An emergent problem with these systems is the lack of privacy. Research on voice scrambling is directed at the development of an inexpensive analogue voice scrambler that would maintain voice quality while permitting only the involved parties to hear the conversation.

In the field of image processing, research is in progress into the most efficient, inexpensive means of transmitting patients' X-ray images from clinic to hospital for immediate diagnosis and treatment. Work is also underway in the development of image compression techniques for the storage and transmission of medical images and X-rays.

The shortage of labourers is a growing problem in Japan and workers are under considerable pressure to put in long hours of overtime. Manufacturers are seeking a solution to this problem in the introduction of robots. Knowing that labour shortage will be an ongoing problem and that robot technology is becoming increasingly sophisticated, Iida believes security robots have great potential.

Researchers in the IS Laboratory's various divisions are working together on the development of a robot capable of providing on-site security and conducting periodic checks.

How to proceed with the research is left to the researchers. Initially an approved research plan is presented to all members of the Intelligent Systems Laboratory. As a periodic check on how the research is progressing researchers must from time to time present progress reports at laboratory-wide seminars. In addition, within each research section there are weekly presentations by researchers on the present status of their work or on related themes. These presentations help the researchers to hone their data collection, interpretation and presentation skills. Foreign researchers are free to use either English or Japanese for their written and oral presentations.

The company encourages researchers to participate in academic meetings. 'The external presentation of research activities encourages researchers in their endeavours and is vital to individual and corporate development', say Isao Masuda, director of the pattern information processing division. With regard to the secrecy issue Hashimoto says, 'We feel there is more benefit to be gained through exposure of the SECOM name in this way in the attraction of top flight researchers than there is to be lost through the exposure of research secrets.' To facilitate participation in international meetings SECOM offers English conversation classes twice a week. These classes are especially useful for those wishing to take advantage of the IS Laboratory's scholarship programme and pursue their studies overseas. In 1990 two researchers were studying abroad, one at the University of Edinburgh and the other at Massachusetts Institute of Technology.

Believing that quality research deserves quality equipment, Iida seldom hesitates to purchase whatever tools the researchers require. The communication link between the research laboratory and Iida's office is short and direct, eliminating many of the problems often associated with decisions on research and development. A longer and/ or more complex chain of communication for project or capital expenditure approval, Iida argues, may result in the transmission of misunderstood or distorted information causing problems for the decision-makers as well as lost opportunity.

While companies outside Japan rely on government funding to cover an average of 60 per cent of their research expenses (an exception to this are German companies which fund about 80 per cent of their own research) Japanese companies rely on their own resources for 98 per cent of their research funding.[6] At SECOM all research is company funded and decisions on expenditure can be made swiftly.

The benefit of this is felt particularly by foreign researchers in Japan. Mark Drumm, a University of California, Santa Cruz, graduate in computer and information sciences, joined SECOM as a researcher in image processing in July 1987 on a two-year contract which has since been extended. Says Drumm:

> Working conditions in SECOM's research facilities are excellent. Researchers feel there are very few limitations placed on them. Equipment and supplies are simply requested of the direct supervisor as we need them and top-of-the-line items are purchased immediately.

The existence of an effective communication network encompassing external researchers, Hashimoto believes, is the key to holding a position on the leading edge of the research frontier. At the research institute, researchers' terminals are connected to a variety of computers within SECOM; in addition, the institute is linked to the Japanese computer network Junet, providing electronic messaging and file exchange with other researchers in a global context. Hashimoto explains:

> It takes about ten years for a research institute to become firmly established. In order to build a respected institute the basic technology has to be firmly rooted. So far we have concentrated on laying the technological foundations within each of the institute's four research divisions. In the near future I envisage a greater emphasis on the implementation of basic research findings.

Iida is pleased with the way the intelligent systems laboratory is developing:

> My goal since the company's establishment has been the enrichment of everyday life through the use of science and technology. The Technical Centre and the Intelligent Systems Laboratory provide the technical support required for SECOM to move toward the establishment of a social systems industry, my ultimate mission as a businessman.

As Iida grows older the question of corporate succession looms large. While he has no intention of stepping down in the immediate future, Iida has been preparing others to lead the corporation by providing opportunities for their involvement in the design of innovative ventures. SECOM's future leaders, like corporate leaders worldwide, must be visionaries. But visionary foresight is not enough. Decisiveness, sensitivity, and the courage to act are also essential. As Iida has demonstrated in the service sector in the Japanese context, visionary, innovative direction and decisive, committed implementation are the key to entrepreneurial success.

12 Making history

DATA BASE SERVICE

'The entrepreneur is a maker of history, but his guide in making it is his judgement of possibilities and not a calculation of certainties.' [1] Iida's innovative strategy has focused on new business development with the intention of creating new services rather than simply improved products within existing services. Market focused, its aim is to change societal values through the introduction of new services based on new concepts. The outcome has been a series of products/services introduced to the market. The commitment to this innovative strategy is the cornerstone of business survival and expansion. What has made many of SECOM's later innovative services possible was one of Iida's earlier innovative concepts – the use of the on-line communication network for security service.

Throughout most of the company's development Iida's primary focus has been the widest possible diffusion of electronic security systems. Concurrently with securing customers Iida's intention was to control as many telecommunications channels as possible. By the end of 1984 SECOM's on-line communication network with two-way interactive service had expanded to accommodate 180,000 clients throughout Japan. Recognizing that the security-related operations and other company business for which the network was used occupied only a small percentage of the system's capacity, Iida wanted SECOM to facilitate more effective information communication required by a variety of existing and potential users. Iida's intention was to create a social network system which would incorporate, in addition to security, database, VAN, audit, and media services. The integration of these services would augment the primary urban infrastructure of transportation, water, power, and sewage and form what could be called the supra-infrastructure necessary for urban amenities.

Since the mid-eighties Iida has ventured into a series of computer-based information service businesses. His first experience in this arena was the provision of a database service. This business was initiated by Iida's encounter with Yoichi Tao. Tao entered the field of computer science as a physics student at Tokyo University. In 1972, following seven years in graduate school, he found employment with the computer company System House and in 1977 established and became president of Seikatsu Kozo Kenkyusho, a research institute for the study of societal life in the future and engaged in software development and system design.

At about this time the newly developed information communication medium videotex was gaining public attention. Videotex refers to any two-way system that permits text and pictures stored in computers to be transmitted via the telephone network and displayed on a television screen. Initially it was often called viewdata but videotex proved to be the more marketable trade name.

NTT introduced videotex in Japan under the name CAPTAINS (Character and Pattern Telephone Access Information Network System) and maintained tight control over it, refusing to make its protocol public. Tao and his company were thus denied access to knowledge of the internal workings of CAPTAINS. Tao felt there was no justification for NTT's secrecy. The telephone lines belonged to the public and the protocol, he thought, should be public as well.

Videotex services were being developed in many parts of the world. For example, in Britain where videotex originated the service is called Prestel, the United States has Comp-U-Serve and The Source, and Canada has Telidon (a television terminal-based interactive information retrieval system). Telidon became the basis for NAPLPS (North American Presentation Level Protocol Syntax), the North American standard. Its protocol was in the public domain and Tao felt it could be adapted by his company for wider application. Capitalizing on this opportunity Tao developed Japanese language software for Telidon and used the system on each floor of the Yokohama Children's Science Centre opened in May 1984 to provide user information. This was the first application of Telidon in Japan and it was well received by users.

Tao wanted to market Telidon software and system know-how to entrepreneurs interested in entering the videotex business. Believing that Telidon had more immediate application possibilities than CAPTAINS and knowing of Iida's recent involvement in the new media business with his establishment of Dai-ni Den Den and CATV Corporations, Tao visited Iida in the fall of 1984 and encouraged him to become involved in videotex. Iida had been interested in videotex and had been studying CAPTAINS and Telidon. Listening to Tao describe the videotex information service business he envisioned based on Telidon Iida knew intuitively that it would succeed.

Iida's response, however, was not what Tao had anticipated. Instead of looking for other people to delve into this business, Iida suggested that Tao do it himself and offered to back him financially. Iida introduced Tao to the principal associates with whom he had established Dai-ni Den Den, a telecommunication corporation challenging the newly privatized NTT – Kazuo Inamori, Chairman of Kyocera, the world's largest manufacturer of IC ceramic packages and a front runner in bioceramics, including artificial bones; and Jiro Ushio, chairman of Ushio Denki (Ushio Electric Incorporated), a fast growing maker of special lamps including halon lamps and xenon lamps. Both of these men were innovators in their respective fields. Both had gone beyond the scope, objectives, goals, and technology of the time to open up new fields.

In January 1985, only months after Iida's first meeting with Tao, the company Videotex Centre was born with an initial capital of 1.2 billion yen. Of this SECOM put up 28 per cent, Tao and his institute 28 per cent, Kyocera and Mitsubishi Corporation 20 per cent each, and the social engineering institute of which Ushio was the president 4 per cent. Tao was president of Videotex Centre, Iida was chairman, and Inamori, Ushio, and Hiromune Minakawa, a former vice-president of Mitsubishi Corporation, were the directors.

At about this time Tao learned that MITI (Ministry of International Trade and Industry) wanted to establish Technomart, a platform for technology information exchange. Soliciting the cooperation of high technology industries, MITI was interested in the formation of a technology information exchange market resting on a computer network encompassing regional high technology centres throughout Japan. The establishment of a new computer network, however, would require a substantial capital outlay and take considerable time to complete. Tao met with MITI officials and suggested that utilization of Videotex Centre's network as the medium for the proposed information exchange would eliminate both these problems. MITI concurred and, under the ministry's guidance, in July 1985 the Technomart Foundation was established with some six hundred high technology corporations participating.

With regard to Videotex Centre's own activities Tao recognized three conceivable focuses for the company's service: families, the general public, and corporations. Using television screens as terminals CAPTAINS had targeted the family market so Tao felt that Videotex

Centre would do best to cultivate the corporate market. The service provided would be functional on a wide variety of computer monitors including NEC, Fujitsu, Toshiba, Hitachi, Mitsubishi, Oki, Ricoh, Sony, and IBM. Iida and the Board of Directors concurred with Tao's decision.

In addition to deciding upon Videotex Centre's target customers it was necessary to determine how its services would be financed. Iida was a strong proponent of the idea that the information users should pay for the service, unlike NTT's CAPTAINS where suppliers pay the costs. Iida elucidates:

> In many instances in the past, regardless of the nature of the information provided, the corporate information sources would pay to be included in the service and such information was provided at little cost or even free to receivers. Because expenses were met by suppliers the database company often had little interest in the quality of its service and its usefulness to receivers.
>
> On the other hand, when it is the receivers who are paying for the service the database company cannot fail to be motivated to provide a high quality information service in order to appeal to subscribers. The service must prove its worth. I thought this would be the preferred way for Videotex Centre to operate. It would guarantee that the company would provide the best service possible. I feel this should be the direction in which the new media information industry heads.
>
> The risks inherent in this approach are far greater than those inherent in the method upon which CAPTAINS operates but business without risk does not provide a worthwhile challenge. Without challenge the entrepreneurial spirit cannot be actualized. The greater the risk the greater also the chance to reap large benefits.
>
> In Japan, paying for information has not yet been firmly established in the societal behaviour although the term 'new information society' or 'new media society' is often used.

Iida's decision on the method through which Videotex Centre's service would be financed was a partial reflection of his frustration with the temporal discrepancy between societal perception and societal behaviour. Iida's aim was to bring about a change in societal perception of the value of information by convincing information providers and users that information and knowledge have attendant costs which are in direct relation to their quality. This approach was similar to that which he had used in the security industry.

Iida was among the first to recognize the need for 'individual' (be they corporate or personal) attentive security services and the first to develop such services. His initial success rested largely on his ability to bring about changes in societal perceptions of the provision of security and the acceptance of the idea that assured security has a price.

In April 1986 Videotex Centre initiated VIEWCOM. VIEWCOM is comprised of two major information services: inter-corporate on-line management information systems and an information service initially targeted on technical information related to the electronics industry.

The first service is for corporate manual management. As corporations grow the number and complexity of the tasks to be completed in order to run effectively increases. To function efficiently individuals require guidance which is frequently provided in the form of manuals, for example, business control manuals, contract manuals, technical manuals, sales manuals, production manuals and so on. These have to be updated as procedures change in accordance with the changing internal and external corporate environment. In recent years corporate adjustment to consumer demand and the changing business environment including technological change has necessitated more frequent changes in corporate manuals with their attendant costs. With VIEWCOM manuals can be updated easily and quickly at relatively low cost and the new format reduces storage problems.

The service Tao developed grew out of his experience with SECOM's business control manuals. Overall improvement in SECOM's corporate system depends on improvement in each of these three supporting pillars: the computer, the communications network, and its personnel. The first two areas can be upgraded through improvements in hardware technology provided that the necessary capital is available. However, bringing about increased effectiveness of human resources is more complex. Training must be ongoing, it cannot be achieved overnight. With the constant additions to and changes in SECOM's products/services employees must be kept up to date with products, services, and technical knowledge. To facilitate this the company developed detailed manuals and internal rules to be followed by employees. In order to keep these up to date a dependable database is essential. Tao developed such a database on VIEWCOM for SECOM. Encouraged by Iida to use SECOM's detailed internal corporate data as an example in his sales pitch, he went on to market this service to other corporations.

VIEWCOM's second component, an information service for the electronics industry, is rapidly gaining popularity. The decision was made to focus this service on the electronics industry because here information providers and users can specify the precise nature of the technology or product, allowing for narrowly focused searches. As well, the amount of information available is immense, making computer searching desirable. Integrated circuits, for example have been used not only in electronic products but in a wide variety of industrial items including automobiles, machines, toys, and watches. In Japan about 500,000 devices such as semi-conductors, condensers, and connectors are on the market with new products being introduced regularly, often as many as 1,000 a month.[2] A product designer wishing to utilize these items must choose from a vast array. In the case of product design a well-classified database can reduce the time and increase the efficiency of product development. The Videotex Centre database, which includes information on electronic devices produced by two hundred manufacturers world-wide, is value added – arranged and edited for the user's convenience.

Tao's ability to recognize the importance of focusing on the segmented market developed from his experience as editor of the personal computer journal *RAM*. Five years in this position taught him that the market for personal computer journals is minutely segmented. Regardless of the various marketing methods employed there is a threshold beyond which circulation cannot be increased. Tao became aware that multi-product small batch production is essential not only in the product sphere but also in the information service sphere. At the same time clearly identified segmented population groups will purchase precise, accurate, and up-to-date information.

VIEWCOM's information service also provides a summary of the latest electronic industry information gleaned from over sixty overseas industry journals and newspapers. In addition, it offers definitions of new technical terms, description and examples of application of the new technology, and analysis of the social and industrial context within which the new technology is developing. The information service includes industrial, trade, regional development, and communication policies related to frontier high technology published by the Ministry of International Trade and Industry (MITI), the Ministry of Posts and Telecommunications, and the Science and Technology Agency.

The above services are all created by Videotex Centre. As well, VIEWCOM offers subscribers access to the Nikkei McGraw-Hill data programme for the electronics industry. By the end of 1990 VIEWCOM's services were utilized by 2,000 corporations. The corporate target is to provide comprehensive information in the areas of finance and material distribution and to provide information on such necessary managerial resources as people, materials, and money. Since August 1986, Videotex Centre has been linked to Secomnet, the nation's largest computer network system. With Iida's strong support Tao is eager to develop further VIEWCOM's services.

VAN SERVICE

Through the process of establishing the computer network system SECOM acquired both information communication technology know-how and a human-machine integrated customer response system. With this high-tech/high-touch 'social systems network' in hand, Iida was ready to actualize SECOM's mission as the architect of a new 'social systems' industry in 1985 when the government controlled telecommunications network industry became privatized and liberalized. With this purpose in mind, in August 1985 Iida established Secomnet Company as a wholly owned subsidiary of SECOM with an initial capital of 300 million yen.

Iida needed a capable individual to launch this new venture. He chose Kei Kimura who had come to SECOM sixteen years earlier from Nippon Denshin Denwa Kosha. Recognizing not only Kimura's accumulated knowledge of communication networks but also his ability to actualize Iida's vision of a communications industry, Iida appointed Kimura president of the new company while he became Chairman of the Board.

This VAN (value added network) system which began in November 1986 had several strengths. Its coverage was nationwide with more access points than any other system. The network had a strong service and follow-up human support system provided by qualified personnel based at each of its seven sales branches, eighteen control centres, and 220 exclusive VAN use access nodes across the country. Secomnet had no attachment to any industrial group, allowing it to remain neutral, free from constraint, and able to offer an integrated access service compatible with a variety of computer systems. For its customers Secomnet provided worry-free network management, line use cost reduction stemming from time-based line usage charges, network security, and round-the-clock uninterrupted service.

The advantages Secomnet offered were recognized widely and when Iida was ready to expand the company with the infusion of an additional 3.3 billion yen less than a year after its establishment, eighteen corporations entered into partnership with Secomnet. Four corporations, Kobe Steel, Kanebo, Digital Computer, and Nihon Dilex, came on stream to facilitate systems building. Seven companies participating in software development are Nippon Jyoho Sangyo, Nippon System Development, Nippon Time Share, Tokai Create, Data Process Consultant, AGS, and Data Application. Seven others joined for the purpose of facilitating market development: Nissho Iwai, Tokyo Kaijo Kasai Hoken, Sanwa Bank, Sumitomo Bank, Taiyo Kobe Bank, Tokai Bank, and Mitsubishi Bank.

IBM also expressed interest in Secomnet. SECOM had been using IBM 3083s as host computers for the security business. To accommodate the VAN service an IBM 3090 was installed. Through these business dealings Iida and Takeo Shiina, president of Japan IBM, came to entertain mutually the desire for cooperation in the form of network sharing. IBM was interested in utilizing Secomnet's access points and Secomnet, in turn, was attracted by IBM's backbone network which provided high speed connections between major points in the system. The result was an agreement between the two companies in 1986 providing for specified cooperative systems usage.

After linking up with IBM, Secomnet further expanded its cooperative systems usage through an agreement with Kyodo VAN, a joint venture established in 1984 by fifty-eight companies of various types in conjunction with CSK (Computer Service Kaisha). Secomnet was attracted by the variety of services Kyodo VAN offered including packet exchange, commodity flow and financial information data exchange, and information processing. Having only forty access points, Kyodo VAN was attracted by Secomnet's 220 access points. In July 1986 the two companies entered into an agreement allowing for synergistic network usage cooperation.

To accommodate expanding computer usage and to facilitate storage and distribution of system components the FS (future structure/full support) Centre was opened in June 1988 in Yokohama. The 14,000 square metre building consists of two six-storey sections, one for computer controlled storage and the other Secomnet's computer centre, linked by a single-storey office unit.

A recent expansion undertaken by Secomnet, in conjunction with the Nippon Sekiyu Gasu Corporation, is the development of a system for the security and computerized consumption monitoring of LPG (liquified petroleum gas) in home and industrial usage. In the area of security, a sensor-identified gas leakage will automatically result in a shut-off of LPG. The data is transmitted to the gas company via the FS Centre by facsimile transmission and emergency procedures are then activated by the company.

In the area of consumption management consumption and amount remaining are recorded automatically at the FS Centre and the data tabulated in relation to delivery and payment information. Based on this data the gas company bills its clients and delivers LPG. This system allows Nippon Sekiyu Gasu to offer its customers a high level of security, and to provide them with a steady supply of LPG, while affording the gas company a highly efficient gas usage and payment database. Increased popularization of this system means a growing clientele for SECOM, especially if the gas company customers were not previously users of SECOM's services.

Through the connection of clients' terminals to SECOM's host network, customers can use the network as if it were their own. This service not only allows clients access to SECOM's general business information data base, it also permits the storage and retrieval of individual client files. Access to the service is nation-wide enabling even clients' distant branches to use SECOM's host computer as if it were their own.

SECOM not only provides the appropriate software package suitable for each customer's specific needs, it also offers a consulting service for the development of the client's network architecture, software development, business strategy system architecture, and system integration. As well, it assists clients in the selection of the appropriate hardware and software necessary for the client's network architecture and is engaged in the sales and maintenance of equipment.

In 1990 Secomnet introduced an economical voice-line network service which links, by a direct dial line, all the nodes of a client's business within Japan including the head office, branches, affiliated corporations, suppliers and distributors, factories, distribution centres, and customers.

COMPUTER SECURITY

Today a wide variety of social institutions including medical, political, educational, financial, transportation, welfare, and taxation institutions, are supported by computer network systems. As dependency on such networks increases, so does the potential havoc to be wrecked on society through malfunction or improper use.

As a security specialist, Iida was perturbed by the vulnerability of computer network systems in Japan. As early as 1974 he was discussing the need for computer security measures with his employees. In 1982 a keynote address by an American computer crime specialist at a SECOM Science and Technology Foundation sponsored symposium heightened public awareness of the need for computer security.

In 1984 Iida chose Yasumichi Ueno to head an investigation into the feasibility of SECOM establishing a computer security business. Ueno, a graduate of Iida's alma mater, had left the Japan Nestlé Corporation and joined Nihon Keibi Hosho in 1974. Ten years later he was the company's personnel manager, a position he had to relinquish in order to head the new project. By this time computer security was emerging as a national concern. Such government agencies as MITI, the Ministry of Posts and Telecommunications, FISC (Financial Information Systems Centre), a foundation affiliated with the Ministry of Finance, and the National Police Agency were developing strategies for the efficient and secure use of computer systems.

The organizational design principal for SECOM's innovative ventures has always been the 'project team', an autonomous ad hoc unit existing outside of the organizational format within which current business operates. Ueno coordinated a project team and examined the proposition. Their conclusion was that SECOM's expansion into the field of computer security was essential but could not be carried out independently. Says Ueno:

> Computer security requires knowledge of computer damage protection, network systems, and financial management. Our company had personnel who could deal adequately with the first two areas but none well versed in the financial management required for system auditing in the computer security business.
>
> The project team concluded that a joint venture would provide the most expedient entry into the field of computer security. The selection of a partner was crucial to our success. Instead of joining forces with a computer manufacturer, we chose NTT to supplement and synthesize our strengths.

SECOM proposed the establishment of a joint venture with NTT in February 1985 at the time when the latter company was becoming privatized. NTT was in favour of the venture from the outset. By the end of 1984 SECOM was utilizing 80,000 private communication circuits, 18 per cent of NTT's total private circuit capacity, and 40,000 public communication circuits, making it NTT's largest customer. Through their long association NTT had come to know and understand SECOM well. Now in the process of privatization, NTT was seeking low risk expansion opportunities. For NTT, SECOM was an ideal partner.

Several meetings were held and in June, following completion of the privatization process, a committee was formed to deal with the details of the establishment of a joint venture. In October 1985 the Nippon Computer Security Corporation (NICOSE) was born. SECOM put up 55 per cent of the 350 million yen investment capital and NTT the remainder. Ueno became president.

NICOSE is involved in four different facets of the computer business: computer system auditing, computer system consultation, computer network analysis, and computer security measures. The year NICOSE was established the government bodies promulgated various computer system safety criteria. In January 1985 MITI designated system auditing criteria and the following December the Ministry of Posts and Telecommunications' advisory body, the Telecommunication Technology Commission, published a report on telecommunication system safety and dependability. The same month FISC announced criteria for safety measures pertaining to computer systems for financial institutions. In January 1986 the National Police Agency's Computer Systems Safety Measures Study Group published a report on achieving computer security. In its newest joint venture SECOM was moving with the times.

Among the four activities in which NICOSE is involved, systems auditing is perceived to be the most important. The first three months of 1986 were devoted to establishing NICOSE's criteria for systems auditing. After studying the criteria issued by the three government bodies, the company arrived at approximately 3,000 system auditing points. NICOSE entered into a technical tie-up with Aoyama Auditing Corporation, Price Waterhouse's Japanese representative and began their system auditing service in April.

There were many corporations which required systems auditing but few had either the necessary internal organizational setting or the trained personnel required to carry it out. In May NICOSE began an instructional programme to prepare future systems auditors for the MITI administered information processing systems auditing examinations to be introduced in the autumn.

Japan was becoming a highly sophisticated information society. The Achilles' heel of such a society was sharply revealed by two events. In May 1985 the international communication circuits of KDD were broken into and the following year the National Railway's communication network was destroyed by Japanese terrorists. Public concern over the vulnerability of computer networks was mounting. NHK (Japan Broadcasting Corporation), in conjunction with SECOM, organized a computer security symposium. Among the participants was John F. Maxfield, a former FBI computer security consultant.

Critical information may be leaked from a number of points in a computer system. One of the areas most vulnerable to computer 'hackers' is the telecommunication circuits, thus telecommunication corporations are involved in the development of cryptographic transmissions. Using the most advanced technology NTT, SECOM's partner in NICOSE, developed FEAL-8 in 1986. NTT claims that it is almost impossible for a potential intruder to decipher FEAL-8 transmissions. A rumour began to circulate in Europe in the spring of 1989 that the code had been cracked. To dispel the fears of users and potential users and quell the rumour, NTT is challenging hackers to break the code within two years for a 'reward' of one million yen. More likely a serious criminal successful in breaking the code will not come forward as the profits to be made from the illegal use of this knowledge are far greater than the reward money.

COMMUNITY-BASED SERVICE

Iida envisaged expansion of his concept of 'security' in the sense of 'peace of mind' and in 1984 chose Tsuneo Komatsuzaki to head a ten-member project team to investigate a number of possibilities. A graduate of Waseda University specializing in electronics and computer communication, Komatsuzaki joined the company in 1978 and worked for two years in the SP Alarm division as a sales engineer. From 1980 through 1983 he was employed in the planning department both at head office and at the Miyazaki branch. As the planning department reports directly to Iida and is located in close proximity to Iida's office, Iida had the opportunity to get to know Komatsuzaki and observe his performance. Komatsuzaki's eagerness to challenge new ideas and to work under his own initiative were characteristics needed for the exploration of possible new ventures.

Computer engineers and marketing specialists comprised the remaining nine members of the project team. Several factors entered into the team's formulation of target business activities for the company. Past diversification had centred mainly on new applications for the SP Alarm system. Komatsuzaki was excited by the team's mandate to look to 'virgin territory' for expansion possibilities. From his work in the planning department Komatsuzaki was aware that SECOM was already moving into the area of medical network service and wished in some way to tie up with this through health-related activities in which it was feasible for the company to become involved.

Whatever ventures were selected, there would be risks. Whether or not these ventures realized a profit initially, their activities had to contribute to the attainment of corporate health. Finally, there was land available in Kohoku New Town, a recently developed suburb north of Yokohama with a growth limit of 300,000. The community's 80,000 inhabitants were mostly young people with substantial disposable income and an appetite for new things. The project team recognized two dominant family compositions within this group. One was the young family with small children and the other was the established, middle-aged family with grown or nearly grown children. Taking all these factors into consideration the team decided on three ventures.

The first was a drop-in child care service. The team recognized the growing need on the part of young, nuclear families for irregular babysitting services. In bedroom communities such as Kohoku New Town grandparents and other family members were seldom at hand to provide occasional care for young children when it was needed. The realization of this venture was an outcome of the extensive network Iida had developed. Over the years Iida had acquired wide ranging connections and it was this resource upon which he was able to capitalize in diversifying his company's social mission.

Iida was acquainted with Noriko Nakamura, founder and director of JAFE, the Japan Association of Female Executives. A former television announcer, Nakamura had found it extremely difficult to make satisfactory babysitting arrangements when her first child was born. Subsequently she resigned from the television station and in 1986 established JAFE Poppins Service to provide in-home babysitters for female workers. Capitalizing on Iida's connection, as a joint venture with JAFE Poppins Service, SECOM established My Poppins in May 1988. Open from 9 a.m. until 10 p.m. 365 days a year, My Poppins is a babysitting centre where, by appointment, children from 6 months to 6 years can be left for an hour or more any day of the week. The service is not available, however, to working women requiring regular day care for their children. About 400 children a month are dropped off at My Poppins, 65 per cent of them in the morning. My Poppins is located on the third floor of SECOM Centre, a five-storey building opened in Kohoku New Town in May 1988.

The top floor of the Centre is occupied by My Fit, a fitness club offering individually tailored health programmes. This was the second venture the project team recommended. The club offers aerobic exercise classes and workouts with exercise machines. Open every day of the year the same hours as My Poppins, it has a membership of 400 representing some 300 families. Users of My Poppins and My Fit must pay an initial sign-up fee and an annual membership fee in addition, in the case of My Poppins, to an hourly user fee. For those who have a SECOM security system in their home, however, the sign-up fee for both services is waived.

The third venture into which SECOM has entered is the establishment of a medical clinic in partnership with a number of doctors. The clinic is housed on the second floor of the SECOM Centre. It is the initial step in SECOM's entry into the provision of medical services.

All three ventures are administered by SECOM 24 (24 symbolizing corporate intent to offer round-the-clock service), a company formed for the purpose in November 1986. At that time the original ten-member project team was dissolved and replaced by a seven-member implementation team. Under the directorship of Komatsuzaki, these were SECOM 24's first employees. The company now has thirty full-time and twelve part-time employees. The latter are six fitness instructors and six child-care workers who cover the peak hours. By the end of 1989 SECOM 24 had not realized a profit although total sales were good. More than profit making ventures, My Poppins, My Fit, and the clinic are perceived as sales tools which help to spread SECOM's name. Says Komatsuzaki,

> In Japanese families it is usual for the wife to make the initial selection and the husband to make the final decision on a household purchase. Exposure of females to SECOM's name will improve the chances of a SECOM service being selected in the future as SECOM continues to expand into the domestic market.

NEW HORIZONS

Recently SECOM's diversification activities using the security communication's network have extended to the travel business through capital investment in the independent, mid-sized Gloria Travel Agency. There are some 9,500 registered travel agencies in Japan, some specializing in domestic travel, some in foreign travel, and some which deal with all types of travel. Recently there has been considerable growth in the latter type. In the first six months of 1989 for example, the Ministry of Transport approved the establishment of thirty-three new agencies, eleven of them as a means of diversification for corporations not formerly associated with the travel industry. Established in 1967, Gloria's total sales for 1988 were about seven billion yen, 90 per cent of which were derived from Japanese travelling abroad. The tie-up with SECOM was undertaken for the purpose of expanding the domestic travel business.

SECOM owned 20 per cent of Gloria's shares in 1989. Iida assigned one SECOM executive to head the three-year development of a computer controlled travel service. Utilizing SECOM's extensive telecommunications network, Gloria's computer capabilities are being upgraded and expanded to provide a faster, more efficient, and extensive service. Iida anticipates that over half of SECOM's 270,000 present individual and corporate customers will take advantage of Gloria's travel service. Through the improvement of clients' computer terminals presently used for security purposes, without leaving the premises clients will be able to purchase tickets for domestic travel.

In 1983 SECOM, in conjunction with Kamei and Fuji Xerox, established Miyagi Network, a CATV corporation in Sendai city in northern Honshu. Six years later the corporation received permission from the Ministry of Posts and Telecommunications to be involved in space cable television broadcasting using satellite communications. Via cable subscribers receive multi-channel selection, clear pictures, and clear sound, without the inconvenience of a space-consuming parabolic antenna. The initial service includes movies, overseas sports and news, music, weather, stock market information, and local events calendar.

In the area of transportation SECOM, in a joint venture with Fuji Heavy Industries, Fuji Vending, and Nippon Kogyo Bank, acquired the Nippon Norin Helicopter Corporation in July 1989. SECOM put up 25 per cent of the required funds. Norin Helicopter has one hundred helicopters (approximately one-tenth of all private helicopters in Japan) and 85 per cent of their present business is in the agricultural pesticide service. The use of helicopters in transportation is expected to expand in Japan and SECOM is already using them to transport its employees for business purposes and to move equipment required for the construction of new security infrastructures.

In response to consumer awareness of the need to be security conscious, to introduce consumers to the products and systems that are available, and to provide a home safety consultation service, SECOM opened a home security shop in December 1988. Since then thirteen more security shops have been opened in Tokyo and Kanagawa prefectures where the immediate home security market is felt to be the most mature. SECOM's home security unit is exhibited in these shops and toys and household products tested in SECOM's laboratory and found safe are offered for sale. These shops are one catalyst in SECOM's regionally focused integrated sales strategy.

With increased information technology, global competition, and the crowding of communication channels, the problem of corporate and personal electronic eavesdropping was inevitable. Ensuring privacy with facsimile information transmitted via telephone is a particular challenge. Iida saw the need to develop facsimile communication data security. With three million facsimile machines throughout Japan problems have begun to emerge particularly in the areas of misdirected transmissions (dialling a wrong number) and 'junk mail'. NICOSE has developed SCRA, an adapter to be used by both sender and receiver. SCRA, which uses identification numbers and scrambler mode, is compatible with the most popular types of FAX machines in use in Japan.

The technology SECOM developed for building security and maintenance is now being applied to large-scale hydroponic gardening. SECOM, together with SECOM Industry, purchased a 13,000 square metre property, formerly a government seedling nursery, on the outskirts of Shiroishi city in Miyagi prefecture and established SECOM High Plant headed by Seiji Yoshida, president of SECOM Industry. After graduating from Kanazawa University Yoshida spent twenty years with TEAC, a major manufacturer of magnetic recording devices. He joined SECOM in 1983 as a member of the Product Development Office of the Technical Centre. In 1986 he was appointed Executive Managing Director of SECOM Industry and two years later he became its president.

SECOM High Plant integrates frontier technology including cultivation technology, environmental control technology, communication technology, new material technology, bio-technology and robot technology. Opened in February 1990, the 1,000 square metre factory situated next to the administrative centre has a computer controlled environment utilizing entirely artificial light which allows for year-round hydroponic plant cultivation and multiple harvests. Still in its experimental stage, SECOM High Plant is presently concentrating on the production of herbs such as peppermint, thyme, and sage. Under normal cultivation these herbs take two months to reach maturity. Grown hydroponically they can be harvested every two weeks. A high value added product, annual sales for 1992 are expected to reach one hundred million yen.

This is the initial step in the development of a system that will allow for the year-round cultivation of vegetables, fruits, and flowers, particularly varieties not native to Japan. Iida intends eventually to market this system globally for use particularly in northern and arid regions where 'natural' vegetable and fruit cultivation is difficult if not impossible. He sees future application for the technology in space colonization.

Environmentally controlled hydroponic cultivation also has application in the production of seedlings and saplings that will then be transferred to a natural growing environment. A reliable supply of quality seedlings and saplings is important not only for agricultural and horticultural production, but also for reforestation, the retardation of desertification, and the greening of the urban environment, towards the rejuvenation of the planet. In accordance with its corporate mission, SECOM has recently engaged in the development of biotechnology which will have application in the control of environmental pollution. For this purpose Iida established Tecno-Bio Company Limited in 1990.

In the area of medical security SECOM has developed and marketed 'My Care', a self-health management system. This product emerged out of the know-how accumulated through information communication and the desire to participate actively in the medical information business. The system makes it possible to check blood pressure, administer electrocardiograms, and analyse urine samples at home. The first two procedures are carried out by sitting for two minutes in a computerized chair, the base of which also serves as a scale, and the third by placing the specimen in a computerized unit. While other sophisticated, high-tech instruments for home measurements are commercially available, what makes 'My Care' an innovative product is the fact that it is connected to SECOM's central computer which is in turn linked to hospital and clinic terminals. Each measurement is visible in the liquid crystal display window for the user's immediate information and tabulated monthly reports are sent to clients. Up to four people can be accommodated by the basic 'My Care' system.

The system is designed to be used under a doctor's guidance and is aimed at the early detection of heart disease, high blood pressure, and diabetes, three common ailments among the adult population in modern Japan, for which, often, there are few symptoms in the early stages. Says Hajime Takahashi of SECOM's Office of Product Planning:

> In today's busy world many Japanese fail to have regular medical check-ups. Even in the corporate environment where check-ups are given at the workplace these are, at best, only annual occurrences. The data they provide is sporadic. Continuous readings are required if significant changes are to be recognised

in their early stages allowing preventative measures to be taken before the onset of a serious medical condition.

Development time for SECOM's products averages about one year but 'My Care' required three years to develop in order to meet the stringent safety standards imposed on medical equipment by the Department of Health and Welfare. Perfection of the system's software was time consuming. Compared with SECOM's home security system 'My Care' required more sophisticated information processing. There could be no mistakes in data transmission and processing.

FAMILY FRONT

Since the company's inception, Iida's energies have been directed almost entirely towards the actualization of his corporate vision. In the early days he would work late into the night making the rounds of his clients' premises to encourage his guards. Later some of his time was spent travelling among the company's branches. Domestically, where possible, he travels by company helicopter but time is always scarce. One senses that there can be little time left for family life but this is a subject Iida relegates to the private domain and prefers not to discuss.

Weekends are his own and often Iida devotes them to family activities. His daughter, Mami, has fond memories of time spent with her father during her childhood and youth.

> My father often spent the weekend with us and we'd go shopping and have dinner out. Weeknights I remember we'd often spend time together talking and playing cards. Sometimes he and I would get into heated discussions of societal issues. For reasons I couldn't understand, he would occasionally get very upset with my point of view but I am my father's daughter – I would never back down. Sometimes I'd end up in tears and he'd disappear. What I remember most was being able to express my true feelings to my father even if I couldn't make him see my perspective.
>
> My father learned self-discipline from his father and from his life experience and has passed the lessons on to his children. When I was little he would reprimand me if, after playing, I complained that I was tired. He taught me that to play was my choice and I must learn to accept the consequences of my choices without complaint. He taught me to be in control of my behaviour, and my spiritual stance.

I feel my father is broadminded, serious about life, honest, and thoughtful. I recently learned that on the occasion of my marriage this spring my father generously donated to an association for the handicapped. I think he did this to express his thankfulness for our family's good fortune both in the past and in the future.

13 Conclusion

The objective of the foregoing study is to give insight into the nature of entrepreneurial activities in the service sector in Japan through the exploration of one example. The role of entrepreneurial vision in corporate development has been documented through the identification of the evolution of SECOM under the orchestration of Makoto Iida.

Disillusioned by his inability to realize his creative potential while working in his father's sake business, Makoto Iida accepted the risk of establishing an innovative, entrepreneurial firm, venturing into a service area hitherto unknown to Japanese clients. Since SECOM's inception Iida has been the grand designer and executor of corporate development. He has focused on the cultivation of a dynamic, market-oriented corporate culture. Externally, the emphasis has been on the creation of value for potential customers. Once the firm was firmly established, but recognizing the limitations of a human-based security service, Iida dramatically restructured his business by incorporating high technology. The machine security that revolutionized the traditional human-based system became an instrument of vision. The provision of a security network to prevent unlawful entry and fire created the opportunity for the business to delve into the provision of a variety of social security services. Iida's innovative decision to offer an on-line security service provided the opportunity to identify societal needs, to capitalize on social changes, and to bring benefit to all the constituent groups with which his business interacts.

No claim is made that the process described in this case history is unique or universal in either the domestic or global context. Some observations, however, may be useful.

Where does Iida fit into the entrepreneurial history of Japan, especially with regard to the legitimacy of entrepreneurship? In the Edo period, towards the end of the feudal society, entrepreneurial activity emerged within the feudal merchant class. Iida's ancestors were among these early entrepreneurs. With the coming of the Meiji era new entrepreneurial ideology developed, strongly influenced by the statesman, Eiichi Shibusawa (1840–1931) and the educator/social reformer, Yukichi Fukuzawa (1835–1905). Incorporating the importance of virtue and morality from Confucianism, the new business ideology practised by the merchant class and samurai-turned-businessmen had, at its core, duty to the country and service in the public interest.

Iida's grandfather, raised in this milieu, developed a business philosophy with honesty the principal tenet. This was passed on to Iida's father who absorbed it so thoroughly that even when the 'obvious' road to postwar survival was via the black market, he studiously avoided all illegal activity. Iida is an astute businessman alert to any new opportunity and he too adheres to a strict code of ethics. While many of his contemporaries have been implicated in political/business scandals – Lockheed, Recruit, and the recent real estate and securities scandals, among others – Iida's reputation has remained unblemished.

It has been pointed out that 'the socioeconomic, particularly occupational, origins of entrepreneurs have been considered in relation to the factors causing entrepreneurial emergence.'[1] Because of the advantages that may accompany them – capital, knowledge, business networks, and the ability to recognize and act upon opportunities – merchants have been perceived as particularly likely to evolve into industrial entrepreneurs.[2] Having merchant class origins, Iida was well situated to emerge as an entrepreneur. He can be labelled as an entrepreneur from the dominant group belonging to the society's mainstream. However, despite his family's merchant background, Iida was without ready capital and a business network.

Iida's business success rests on the cultivation of a corporate ethos which makes possible the operation of organization dynamics in the direction of the ideal goal which is itself changing in accordance with societal changes. The innovative corporation will have a promising future, not because the entrepreneur contemplates progress, but because he acts to make it happen. Iida has been building an innovative organization – an organization that resists stagnation.

Iida's propensity to challenge the status quo has been the driving force for his business development. His insistence on prepayment for service at a time when such a custom was unheard of, the continuous improvement of service and products, backed by research and development, and his ongoing improvement of the infrastructure including the control centre and computer controlled communication network derived from his belief that whatever exists now will become obsolete. Where is this thinking rooted? Iida pinpoints his experience of World War II as the source of his commitment to challenge established values.

The war influenced the formation of the guiding principles of many Japanese businessmen. While Iida's experience of the war occurred in his childhood and youth, many others were profoundly affected by their military service. Many of Japan's prominent postwar businessmen are World War II survivors who feel, in the light of the country's enormous losses, that their lives were spared for a purpose and they must use their gift of life for societal benefit. Koichi Tsukamoto of Wacoal, Japan's largest manufacturer of women's underwear; Kihachiro Onitsuka of ASICS; and Isao Nakauchi of Daiei, the country's largest supermarket chain are examples of entrepreneurs who fit into this category.

Japan's overwhelming defeat in the war became the driving force for the nation's postwar economic development. The country's energy was focused on the reconstruction of the nation and before long not only was the country rebuilt, it had become one of the leading figures on the economic scene. Japan's prosperity has been achieved at a tremendous cost to its citizens, especially in terms of working conditions, living conditions, family relationships, and social welfare. Indeed, Japanese managers and workers alike view business very seriously and devote long hours to their work and work-related activities. This drive to succeed derives in part from a feeling of collective insecurity rooted in the knowledge that Japan has limited natural resources. Contributing to this drive as well is the knowledge, shared by those who lived through the war, of total and absolute defeat of a country never before invaded and nurturing a long-held belief in its own invincibility.

Dramatic events in the lives of individuals, a death in the family or a critical illness, for example, have been known to be the impetus for their determination to enter into a new business venture or make marked changes in an existing one. In Japan, contemporary senior managers have a shared experience of the war. In many cases this resulted in entrepreneurial activity, however, not all were successful. Japanese businessmen have, for generations, been fascinated by the art of war and have studied military strategy for its application to business practice. The similarity between military operations and business activity is one of strategic management, both demand efficient leadership and realistic objectives.[3] Legendary medieval battles, the Russo-Japanese war, World War II, the writings of Sun Tzu, a sixth-century Chinese general and military theorist, and of Clausewitz, a German military strategist, are learning aids. Even though the number of businessmen having first-hand knowledge of war is declining, warfare continues to provide lessons applicable to the business context.

While Iida, like other successful entrepreneurs around the world, can be labelled as an innovator, what particular behaviour made him successful? After a lengthy search for the 'right' venture, Iida was exposed to the concept of professional security. Although this service was unknown in Japan, it was well established in the global perspective: Iida was not its originator. Iida's success lay in his ability to translate new ideas into commercially viable products and services.

While the receptivity of the social environment in which Iida established his company had not previously been cultivated, Iida was able to capitalize on widely known major events like the Olympic Games to establish the reputation of his company. Government response to the emergence of a private security firm was mixed. The existence of such a corporation, made possible by foreign capital, created an uneasiness which led eventually to the development of a privately owned competitor financed and staffed largely by former civil servants. Numerous additional private security firms emerged over the years. The competitive environment provided a positive influence on SECOM's development. SECOM's success is the outcome of Iida's indefatigable, ongoing search for the answer to the question: 'How can we serve our customers better?' His answer derives from the market-focused grand design he continues to paint and repaint.

At the outset, Iida made a conscious effort to avoid exposure to foreign security practices in order to be free to build a business that would cater directly to Japanese needs. Throughout his life, however, Iida's eyes have been open to the outside world and he has been receptive to western flavours. Iida's interest in American football in university, his early contact with Ligue Internationale des Sociétés de Surveillance, the incorporation of European security symbols into the first corporate logo, the decision to change the company name to an English name (SECOM – Security Communication), the use of English terminology in the naming of key corporate facilities such as the research centre and the staff training centre, the exposure of company personnel to western ideas and technologies through the study abroad programme and the incorporation of foreign scientists into SECOM's staff, and the use of foreigners and foreign scenes in brochures and catalogues directed at the domestic market are all indications of Iida's fascination with the West.

Innovation and the diffusion of ideas and technologies creates dynamic change on the global scale. In the world of business the acquisition of new ideas through this process is often the key to survival and growth. Scholars have pointed out that since opening the nation to the outside world in the mid-nineteenth century Japanese entrepreneurs have eagerly adopted and adapted western ideas and technologies. Like their German counterparts Japanese have borrowed extensively from Britain and France but they have also turned to Germany, the United States, and other countries for know-how and inspiration.[4] This tradition is alive in the contemporary business world.

Konosuke Matsushita established his own business in 1917 after modifying a domestic electrical product while in the employ of the Osaka Electric Light Company. His major breakthrough came in the early 1950s when he went abroad to study American and European business scenes including markets, products, and production methods. With his newly acquired knowledge he effectively integrated research and development, subcontract network production, and price conscious marketing strategy to create one of the world's largest consumer electronic manufacturers.[5]

When Tadao Yoshida, founder and president of YKK, compared his early zippers to American products he was ashamed and embarrassed by their inferiority. This motivated him to move into that vertical integration and mechanization that enabled YKK to become the world's biggest zipper manufacturer.[6] Similarly, Kihachiro Onitsuka of ASICS Corporation was exposed to foreign products when he attended his first Olympic Games and vowed to produce shoes that would be competitive in the international scene.[7]

While these examples show that western products have been the catalyst for product improvement, in other cases the corporate design has been studied and, where feasible, integrated into the Japanese corporate setting, or perceived as an example of what could be improved upon. Toyota's innovative *kanban* and 'just-in-time' production system, which links marketing, production, and customer, was triggered by the shortcomings of Henry Ford's pioneering but inflexible mass production system.[8] Interest in the outside world continues to be a common characteristic of Japanese entrepreneurs and Iida is no exception. The transfer of ideas obtained from the outside to new products, services, and business opportunities at SECOM is in accordance with Japanese entrepreneurial tradition.

Innovative ventures in the manufacturing sector appear to be fuelled by technology-based product innovation. In the case of Sony and Honda for example, the founders were engineers well versed in product technology. Iida, on the other hand, lacked technical knowledge. His entrepreneurial success was rooted in his ability to design a security system, understood in the broadest sense, integrating people and machines. In SECOM's case, as documented in this study, the key to entrepreneurial success was design-focused vision actualization.

In implementing his vision, Iida was fortunate in having Toda's support and dedication from the beginning. While Iida occupied the limelight, Toda's diligence behind the scenes was indispensable for SECOM's success. Entrepreneurial success stories usually focus on the key player, but often there is a system of 'twin management' in place, with one player in the foreground, the other in the background, balanced partners in corporate success.

At Honda Motors Soichiro Honda focused his expertise on the technological side of the business, while his partner, Takeo Fujisawa, a gifted manager particularly skilled in labour relations, finance, and sales, took on the lion's share of factory management.[9] Likewise Masaru Ibuka, the founder of Tokyo Telecommunications Engineering Corporation, later to become Sony Corporation, and Akio Morita were partners in corporate development.[10] The role of team management in entrepreneurship, a subject only touched on in this study, offers yet another avenue for entrepreneurial understanding.

Despite the partnership with Toda, Iida's leadership style is autocratic. There is perhaps no characteristic of Japanese management more widely discussed in the West than consensus decision making. In fact, however, many of Japan's most successful businessmen are autocrats, either overtly or covertly. Inamori of Kyocera, Onitsuka of ASICS, Ichiro Isoda of Sumitomo Bank, Hisao Tsubouchi of the Kurushimadock Group, Isao Nakauchi of Daiei Inc., Toshiaki and Seiji Tsutsumi of Seibu, Takami Takahashi of Minebea, Seiuemon Inaba of Fanuc, and Keizo Saji of Suntory are all well-known autocratic entrepreneurs.[11]

At SECOM Iida makes all the critical decisions. While he consults his staff, he knows in advance the direction in which he wishes to proceed. Within the future-oriented, strategic guidelines, he provides opportunities for workers willing to challenge new ventures. In certain cases, for example in adopting SECOM as the new corporate name, he proceeds slowly, allowing time for employee attitudes to align themselves with the new vision.

What distinguishes Iida from other senior decision-makers in Japanese corporations is his management style in the foreign arena. After carefully choosing his overseas business partners he entrusts the foreign venture to them. This has become particularly true in the case of SECOM's expansion into the American market.

Corporate strength is determined by the sum of the strength of a company's employees. Successful mobilization of this strength for the attainment of corporate vision requires the commitment and dedication of the management body. Iida had to recruit management personnel from outside the company, a problem shared with other new, expanding corporations. He was blessed with an uncanny ability first of all to find and then to attract the appropriate personnel to implement his ideas and bring them to fruition as constituent parts of his grand design. Of necessity, early recruitment was from among his acquaintances, but later his catchment area widened to include organizations within both the public and private sectors. As identified in this study, the movement of personnel from one company to another is a reality in the contemporary business world in Japan where lifetime employment has been perceived as a distinctive characteristic.

Iida is frequently asked to speak on the key to corporate success. His message is simple: People fail because they stop too soon. Indeed, it has been persistence, a characteristic of successful entrepreneurs world-wide, that has enabled Iida to propel SECOM to the position it occupies today. From his initial insistence on prepayment for the service he proposed to his insistence on the meeting of his financial terms in the international arena, Iida's confidence in SECOM's service has been demonstrated through his persistence in implementing business strategy.

Once the company was launched and enjoying a degree of success, capital, human resources, and the physical infrastructure materialized relatively easily. Their absence never threatened the company's development and their procurement, while of course important, was not the most critical factor in corporate success. Of greatest significance for SECOM's development has been Iida's ability to design the company's future. The stamp of personal leadership has been the necessary energy to build and lead the organization to its present state.

Iida's dilemma has been to cultivate, among his employees, an innovative and challenging spirit which rejects existing values and fossilized business behaviour and seeks to replace them with new challenges firmly placed in the corporate strategy for the future. At the same time he has had to create a culture which would enforce the rigid rules and procedures necessary for the well-structured operation of a security-centred social system industry. In his effort to transform his personal corporate vision into a progressive 'social-system' security firm it has been necessary for Makoto Iida to reconcile the paradoxi-

Makoto Iida (right) and the author. 1991.
10 ~~Iida receiving the Mainichi newspaper industrialist award, 1982.~~

cal forces of corporate growth and entrepreneurial change. This
continues to be a dynamic dialectic process, on one hand balancing
aspects of organizational stability (tight business focus and organizational integrity and cohesion) and on the other hand implementing
a future-focused vision (entrepreneurial spirit, organizational flexibility, and 'hands-on' top management). It has been stressed throughout this study that the binding force which has made possible the
effective operation of an innovative, dynamic but steady corporate
endeavour has been Iida's ability to formulate and, through the astute
utilization of resources, actualize his entrepreneurial vision.

Although the phenomenon of entrepreneurship is extremely widespread each entrepreneurial phenomenon has its own distinctive patterns and traditions. This fact makes it difficult to find a single method adequate for the study of entrepreneurship in general. This study, focusing on a living entrepreneur in the service sector in present day Japan, seeks to identify the reality of contemporary entrepreneurial behaviour in the evolutionary framework. It is hoped that the description offered herein will provide useful data for the formulation of a unifying framework for entrepreneurial research in general and in the Japanese context in particular.

Appendix: A social scientist's summary of entrepreneurship

In the past, there have been many different views on the meaning of the term 'entrepreneur'. As S.M. Kanbur said, 'entrepreneurship is the phenomenon which is most emphasized yet least understood by economists'.[1]

The word entrepreneur, appearing first in the French language, referred to explorers, adventurers, government infrastructure contractors, architects and agricultural contractors. Later, it was extended to include proponents of industry as risk taking or risk calculating capitalists. In the English language the term, as well as the concept, was less familiar. Throughout the development of early English classical economic theory, entrepreneurship did not have a prominent place as the theory focused on the gains from exchange and the division of income among capital, labour and land in a state of static equilibrium. However, entrepreneurship, perceived as theoretically unmeasurable qualities, came to be recognized as a factor in economic performance on a par with, and often superior to, other factors by such theorists of the institutional school as Frederick B. Hawley and John R. Commons and later by Joseph A. Schumpeter.[2]

Entrepreneurship has meant different things to the economist, the anthropologist, the sociologist, and the manager. The economist Joseph Schumpeter was instrumental in the formulation of the view that defined entrepreneurship exclusively in terms of non-routine, creative activities. Schumpeter's view gained prominence over what had been the dominant view of entrepreneurship, namely that the entrepreneur was a passive agent reacting to external factors in his guidance of the firm's development. Schumpeter developed his concept of the entrepreneur as innovator into a model of economic development.[3]

The sociologist Talcott Parsons, taking the Schumpeterian view as his departure point, developed the concept that the entrepreneurial function is a subsystem of the economy which he viewed both as a unit within the larger social system and as a social system in its own right. For Parsons, entrepreneurship is a system of action integrating and coordinating other functions prerequisite to the larger system of which it is a component. Parson subscribes to the view that entrepreneurship need not be confined to the economy; it can account equally well for the occurrence and outcome of changes in systems related to the economy.[4]

Anthropology traditionally concerns itself with innovation (the distinctive feature of entrepreneurship) initially in a primitive context, contributing to the appreciation of entrepreneurship not necessarily in business terms but, rather, in terms of the process of cultural change. Harry G. Barnett's anthropological study concerning the history of innovation offered a rounded discussion of cultural change. He attempted 'to formulate a general theory of the nature of innovation and to analyze the conditions for, and immediate social consequences of, the appearance of novel ideas'.[5] Taking the Schumpeterian view as his starting point, he works backward from it identifying the mental processes related to innovation and accounting for them in cultural terms, thus providing additional interdisciplinary support for the study of entrepreneurship.

The economic historian Arthur H. Cole views entrepreneurship as a dynamic process within the constantly changing business environment. He defines the entrepreneurial function as 'the purposeful activity (including the integrated sequence of decisions) of an individual or groups of associated individuals, undertaken to initiate, maintain, or aggrandize a profit-oriented business unit for the production or distribution of economic goods and services'.[6] Cole sees decision making as the necessary condition in entrepreneurship and innovation as a sufficient condition but not a necessary one.

Chester I. Bernard, a practising senior manager, and Herbert A. Simon, a political scientist with expertise in public administration, both focus on a theory of organizations and in so doing formulate similar definitions of the entrepreneurial function. 'Entrepreneurship is treated by them as residing in that aspect of the functioning system which has to do with maintaining "coordination".'[7]

More recently it has been shown that 'entrepreneurship is essential in the strategic management process providing new content to the strategy of a firm through the identification of opportunities.'[8] As well, the phenomenon has been identified as the engine for growth in the emerging entrepreneurial society. Henry Mintzberg, Michael Porter, Peter Drucker, and Alfred D. Chandler Jr are among those who have contributed directly or indirectly to the understanding of the process of change in society in general and in economic development and entrepreneurship in particular.[9]

Notes

1 INTRODUCTION

1 Recent studies of Japanese entrepreneurship and/or business behaviour in the manufacturing sector include S. Sanders, *Honda: The Man and His Machines*, Boston, Little, Brown and Company, 1975; T. Sakiya, *Honda Motor: The Men, The Management, The Machines*, Tokyo, Kodansha International, 1982; B.J. Rae, *Nissan/Datsun: A History of Nissan Motor Corporation in U.S.A., 1960–1980*, New York, McGraw-Hill, 1982; P. Wickens, *The Road to Nissan – Flexibility, Quality, Teamwork*, London, Macmillan, 1987; W.M. Fruin, *Kikkoman: Company, Clan & Community*, Cambridge, Harvard University Press, 1983; A. Morita with E.M. Reingold and M. Shimomura, *Made in Japan: Akio Morita & Sony*, New York, Dutton Inc., 1986; K. Kamioka, *Japanese Business Pioneers*, Union City, California, Heian International, 1988; T. Yamashita, *The Panasonic Way: From a Chief Executive's Desk*, F. Baldwin (trans.), Tokyo, Kodansha International, 1989. A discussion of the emergence of managerial enterprise in Japan is contained in K. Kobayashi and H. Morikawa (eds), *Development of Managerial Enterprise*, Tokyo, University of Tokyo Press, 1986. A discussion of the importance of innovation in Japanese management is presented in K. Urabe, J. Child, and T. Kagono (eds), *Innovation and Management: International Comparisons*, Berlin, Walter de Gruyter, 1988.
2 'Japan's Smaller Companies: Happy Mediums', *The Economist*, 27 October 1990, p. 73.
3 C.J. McMillan, *Services: Japan's 21st Century Challenge*, Toronto, The Canada–Japan Trade Council, March 1991, p. 11.
4 Hideki Yoshihara observes that 'It is commonly held that Japanese management is bottom-up management. Recently, evolutionary management has gradually become popular among researchers and practitioners, but the strategic management of Japanese companies has remained largely unnoticed. Judging by available discussion and writing, Japanese companies would appear to lack strategic management.' H. Yoshihara, 'Dynamic Synergy and Top Management Leadership: Strategic Innovation in Japanese Companies', in K. Urabe, J. Child, and T. Kagono (eds), op. cit., p. 61.

5 This phrase was coined by V. Gordon Childe in his classic book on social anthropology, *Man Makes Himself*, New York, New American Library, 1961.

6 G.G. Meredith, R.E. Nelson, and P.A. Neck, *The Practice of Entrepreneurship*, Geneva, International Labour Office, 1982, pp. 3–12.

7 In the broadest sense, the entrepreneurial spirit may be found within employees at an operating level. For an exploration of this idea, see A. Stewart, *Team Entrepreneurship*, Newbury Park, California, Sage, 1989.

8 R. Dubos, *So Human an Animal*, New York, Charles Scribner, 1969, p. 238.

9 G. Pinchot III, *Intrapreneuring*, New York, Harper & Row, 1985, pp. 39–40.

10 H. Tanaka, *Personality in Industry: The Human Side of a Japanese Enterprise*, London, Pinter, 1988. A soft cover edition was published by the University of Pennsylvania Press under the title *The Human Side of Japanese Enterprise*, 1988.

11 A recent study written from this perspective and containing many concrete examples is B.B. Tregoe, J.W. Zimmerman, R.A. Smith, and P.M. Tobia, *Vision in Action: Putting a Winning Strategy to Work*, New York, Simon and Schuster, 1989.

12 While new companies are continually emerging bankruptcy is driving others from the corporate scene with similar regularity. According to statistics compiled by Teikoku Data Bank Ltd in 1984 20,363 Japanese companies went bankrupt.

13 The three other corporations are Marui, Dai Nippon Printing, and Toppan Printing. Y. Sakaguchi, *SECOM: Iida Makoto no Joho Senryaku* (SECOM: Makoto Iida's Information Strategy), Tokyo, Paru Shuppan, 1986, p. 5.

14 T. Momose, 'SECOM's Corporate Miracle Alive and Well', *Tokyo Business Today*, July 1989, pp. 48–9.

15 P.S. Christensen, *Strategy, Opportunity Identification, and Entrepreneurship: A Study of the Entrepreneurial Opportunity Identification Process*, Aarhus, Institute of Management, University of Aarhus, 1989, p. 21.

16 I.H. Ansoff, *The New Corporate Strategy*, New York, John Wiley & Sons, 1988.

17 M.A. Maidique and R.H. Hayes, 'The Art of High-Technology Management', *Sloan Management Review*, Winter 1984, vol. 25, no. 2, pp. 17–31.

18 G.L.S. Shackle, foreword to R.F. Hébert and A.N. Link, *The Entrepreneur: Mainstream Views and Radical Critiques*, New York, Praeger, 1982, p. vii.

2 EARLY YEARS

1 F. Tsubouchi, *Akari no Kodogu* (Antique Lighting Equipment), Tokyo, Kogei Shuppan, 1987, pp. 23–7.

2 T. Kamata (ed.), *Shuhan Showa-shi* (The History of Sake Sales in the Showa Period), Nishinomiya, Shuhan Showa-shi Hakkan Iinkai, 1985, p. 15.

3 For an overview of the range of entrepreneurial activity in Meiji Japan, supported by concrete examples see J. Hirschmeier, *The Origins of*

Entrepreneurship in Meiji Japan, Cambridge, Harvard University Press, 1964. For a discussion of the role of the Meiji government in limiting and in facilitating entrepreneurship see Y. Horie, 'Modern Entrepreneurship in Meiji Japan', in W.W. Lockwood (ed.), *The State and Economic Enterprise in Japan*, Princeton, Princeton University Press, 1965; W.W. Lockwood, *The Economic Development of Japan: Growth and Structural Change 1868–1938*, Princeton, Princeton University Press, 1954; K. Yamamura, 'A Re-examination of Entrepreneurship in Meiji Japan (1868–1912)', *Economic History Review*, 1968, Series 2, 21, pp. 144–58.

4 P.H. Wilken, *Entrepreneurship: A Comparative and Historical Study*, Norwood, New Jersey, Albex, 1979, p. 171.

5 ibid., pp. 171–2.

6 E. Shibusawa, *Jitsugyo Koen* (Lectures on Enterprise), Vol. I, Tokyo, 1913, p. 199. Quoted in J. Hirschmeier, op. cit., p. 172; (today, *kigyoka* and *jitsugyoka* are used interchangeably to mean entrepreneur).

7 R. Dubos, *So Human an Animal*, New York, Charles Scribner, 1969, p. 135.

8 H. Edoin, *The Night Tokyo Burned*, New York, St. Martin's Press, 1984, p. 35.

9 Opened in 1887 in Tokyo, Gakushuin University was originally an educational institution for children of the imperial family and the nobility. In 1947, with the abolishment of the peerage system, it became a private school open to the general public.

10 Football was introduced to Japan in 1934 by Paul F. Rusch, an American teaching economics and English at Rikkyo University (known in English as St Paul's University) in Tokyo. The same year Waseda, Meiji, and Kansai universities, following Rikkyo's lead, established football teams. In 1935 Keio and Hosei universities followed suit with Nippon, Doshisha, and Kwansei Gakuin universities doing likewise in 1940. With Japan's entry into World War II the use of the term 'American Football' was banned along with other western sports terminology and the game became known as *gaikyu* or 'foreign ball' although it was seldom played during the war years. In 1947 Kyoto University became the first university in the postwar era to establish a football team in western Japan while, in 1956, Gakushuin was the first in eastern Japan. Today 210 university football teams, 110 high school and sixty company teams throughout Japan belong to the American Football Association of Japan. The sport encompasses some 50,000 amateur players. This information was provided by Mitsuaki Higo of the American Football Association of Japan.

11 First played commercially in Nagoya in 1948, *pachinko* has become Japan's most popular pinball game. Some ten million people, it is estimated, play the game regularly. Prizes such as food, books, watches, and pocket calculators can be exchanged for money if the winner so desires.

12 C. Nakane, *Japanese Society*, Harmondsworth, Penguin Books, 1970, pp. 27–8.

13 T. Kamata (ed.), op. cit., p. 169.

14 M. Kurz, 'Entrepreneurial Activity in a Complex Economy', in J. Ronen (ed.), *Entrepreneurship*, Lexington, Mass., Lexington Books, 1983, p. 298.

15 Keisatsucho Keisatsushi Hensan Iinkai, *Sengo Keisatsu-shi* (Postwar Police History), Tokyo, Keisatsu Kyokai, 1977, p. 319.
16 This information was provided by the Alcoholic Beverage Tax Division of the National Tax Administration.

3 BUSINESS EXPERIENCE

1 For a discussion of new employee training programmes in major Japanese corporations, see H. Tanaka, 'Japanese Method of Preparing Today's Graduate to Become Tomorrow's Manager', *Personnel Journal*, 1980, vol. 59, no. 2, pp. 109–11; H. Tanaka, 'New Employee Education in Japan', *Personnel Journal*, 1981, vol. 60, no. 1, pp. 51–3.
2 The migration figure was supplied by the Consular and Migration Policy Division of the Ministry of Foreign Affairs, Japan.
3 Toda understands that the officer who fired the fatal shot later converted to Catholicism.
4 For a discussion of entrepreneurial opportunity identification see P.S. Christensen, *Strategy, Opportunity Identification and Entrepreneurship: A Study of the Entrepreneurial Opportunity Identification Process*, Aarhus, Institute of Management, University of Aarhus, 1989, pp. 34–54.
5 E.M. Rogers, *Diffusion of Innovations*, New York, Free Press, 1962, p. 13. The process through which an individual adopts a new idea, characteristics of the innovation that affect its rate of adoption, and the success and failure of innovations are some of the issues discussed by Rogers.
6 G.L.S. Shackle, foreword to R.F. Hébert and A.N. Link, *The Entrepreneur: Mainstream Views and Radical Critiques*, New York, Praeger, 1982, p. vii.
7 'General MacArthur's letter to Prime Minister Yoshida concerning police reform and reorganization of the Justice Ministry, September 16, 1947', reprinted in Keisatsucho Keisatsushi Hensan Iinkai, *Sengo Keisatsu-shi* (Postwar Police History), Tokyo, Keisatsu Kyokai, 1977, p. 154.
8 'Police Reorganization Plan', abstracted from a letter from Commander-in-Chief, Far East Command to Prime Minister Yoshida, 16 September 1947, reprinted in *Sengo Keisatsu-shi*, op. cit., pp. 159–62.
9 *Sengo Keisatsu-shi*, op. cit., p. 1198.
10 ibid., pp. 1198–9.
11 D.C. McClelland, 'Achievement Motivation Can Be Developed', *Harvard Business Review*, Nov.–Dec., 1965, p. 7.
12 Joshua Ronen, citing several authors, points out this fact in 'Some Insights into the Entrepreneurial Process', in J. Ronen (ed.), *Entrepreneurship*, Lexington, Mass., Lexington Books, 1983, p. 161.

4 ORIGINATING AN INDUSTRY

1 Japan is considered to be among the world's safest countries. The nation's crime rate is among the lowest in the industrial world. In 1985 there were 1,300 major crimes per 100,000 population in Japan compared to 5,200 in the United States, 6,900 in West Germany and 65,000 in France. In the same year 64 per cent of Japanese offenders were arrested while the

apprehension rate in the United States, West Germany, and France was 21 per cent, 47 per cent and 40 per cent respectively. The estimated population per police officer in Japan in the same year was 556 compared to 361 in the United States, 317 in West Germany, and 276 in France. The estimated area covered per police officer in Japan is 1.7 square kilometres, a little more than in West Germany where it is 1.3 kilometres, but less than France's 2.8 square kilometres, and considerably less than in the United States where it is 14.1 square kilometres (Keiji Seidaku Jitsumu Kenkyu-Kai, *Hanzai Hakusho no Pointo* (Highlights of the White Paper on Crime), Japan Ministry of Finance, 1989, p. 7). To keep society safe in a country where great value is placed on safety and security, over the centuries public service institutions and internal cultural controls have been at work to maintain public security. A recent trend has been the emergence of private security companies. By 1989 nearly 5,000 such companies existed while prior to 1962 there were none.

2 Keisatsucho Keisatsushi Hensan Iinkai, *Sengo Keisatsu-shi* (Postwar Police History), Tokyo, Keisatsu Kyokai, 1977, pp. 688–94.

3 ibid., pp. 927–9.

4 These figures were provided by the public relations office of the Tokyo Metropolitan Police Board.

5 *Sengo Keisatsu-shi*, op. cit., pp. 816–17; According to *1983 Keisatsu Hakusho* (1983 White Paper on Police Services), Tokyo, Ministry of Finance, 1983, p. 132, 686,000 households had been designated as crime prevention communication points by the end of 1982.

6 *Sengo Keisatsu-shi*, op. cit., pp. 520–2.

7 For a discussion of innovative compatibility, see E.M. Rogers, *Diffusion of Innovations*, New York, Free Press, 1962, pp. 126–30.

8 It is common in Japan to ask a person's age soon after being introduced as age is an important factor in performance evaluation. As well, the ages of the participants play an important role in establishing relative degrees of respect in oral and behaviour communications and directly affect the language employed.

9 H. Takeuchi, *Showa Keizai-shi* (Economic History of the Showa Era), Tokyo, Chikuma Shobo, 1988, pp. 158–73.

10 Established in 1986, Jastic Kansai is a joint venture between SECOM and Kobe Steel. Employing personnel freed when the steel industry went into decline, Jastic Kansai provides round-the-clock security guards to sixty clients in the Osaka area.

11 *Sengo Keisatsu-shi*, op. cit., p. 1225.

12 ibid., p. 734.

13 ibid., pp. 1200–1.

14 ibid., pp. 1199–200.

15 ibid., p. 1200.

16 For a discussion of various aspects of labour disputes in the Japanese setting, see T. Hanami, *Rodo Sogi* (Labour Disputes), Tokyo, Nihon Keizai Shimbun, 1976.

17 This information was supplied by the public relations office of the Imperial Hotel.

18 A career diplomat who served as Ambassador to Italy (1930–2) and great Britain (1936–9), Shigeru Yoshida became the most famous politician of

postwar Japan. As head of the Liberal Party he served as Prime Minister five times between May 1946 and December 1954. The 'Yoshida Era' saw Japan's transition from alien control to restoration of sovereignty. A new constitution was promulgated, land reforms instituted, and the occupation ended.

19 J. Murai, *Arigato No Kokoro* (Thank You From the Heart), Tokyo, Mikasa Shobo, 1988, p. 106.

20 ibid.

21 ibid., p. 107.

22 ibid., pp. 111–12.

23 After graduating from Tokyo University, Daigoro Yasukawa together with his father, a well-known coal mine developer, and elder brother established Yasukawa Electric Manufacturing Company Limited in 1915 to manufacture motors. In 1936 he became company president. After World War II he was director of the Coal Agency for a time and in 1956 became the first director of Japan's Atomic Energy Research Institute.

24 This information was conveyed in a letter written by Naruhiko Ueda, a director of Sogo Keibi Hosho, on 28 August 1989.

25 Today, Sogo Keibi Hosho is headed by Tsuneo Murai, son of Jun Murai. Tsuneo Murai edited *High-tech Shokuryo Senryaku* (High-tech Food Strategy), Tokyo, Diamond-sha, 1986 and *High-tech Jyoho Senryaku* (High-tech Information Strategy), Tokyo, Diamond-sha, 1988.

26 25 April 1989, *Keibi Hosho Shimbun*.

27 P.H. Wilken, *Entrepreneurship: A Comparative and Historical Study*, Norwood, New Jersey, Ablex, 1979, pp. 67–8.

5 TECHNOLOGY-BASED INNOVATION

1 For a discussion of the ability to doubt, see J.E. Tropman and G. Morningstar, *Entrepreneurial Systems for the 1990s: Their Creation, Structure, and Management*, New York, Quorum Books, 1989, pp. 198–9.

2 Keisatsucho Keisatsushi Hensan Iinkai, *Sengo Keisatsu-shi* (Postwar Police History), Tokyo, Keisatsu Kyokai, 1977, pp. 664–6.

3 For a discussion of the impact of robotization on the worksite see H. Tanaka, 'Human Implications of Robotization in the Worksite: The Japanese Experience', *Robotics*, 1985, vol. 1, no. 3, pp. 143–53.

4 *Sengo Keisatsu-shi*, op. cit., pp. 811–12.

5 S. Ito (ed.), *Toshi to Hanzai* (The City and Crime), Tokyo, Toyo Keizai-sha, 1982, p. 42.

6 For a discussion of the concept of enterprise differentiation see W. Alderson, *Dynamic Marketing Behaviour*, Homewood, Illinois, Irwin, 1965.

7 There has been discussion for some time now in political circles of the need for a law to protect and keep secret certain categories of information. In 1985 the ruling Liberal Democratic party brought before the National Diet a bill that, if passed, would have achieved this. Opposition from the other parties was strong on the grounds that criteria for determining what qualified as national secrets were vague and that given the power of such a law the government might cast its net too broadly, severely inhibiting

public access to information. With the bill's defeat the LDP began work on a new bill which is expected to come before parliament in the near future.

8 In 1987, membership in *Kyokusa Boryoku Shudan* totalled approximately 35,000. For a summary of the terrorist activities of these groups and police response, see Japan National Police Agency, *1988 Keisatsu Hakusho* (1988 White Paper on Policing), Tokyo, National Police Agency, 1988, pp. 9–85.

9 *Sengo Keisatsu-shi*, op. cit., p. 835.

10 N. Nagayama, *Muchi no Namida* (Tears of Ignorance), Tokyo, Godo Shuppan, 1970 and *Jinmin o Wasureta Kanariya* (The Canary Who Divorced Itself from the Public), Tokyo, Henkyosha, 1970. The seventh of eight children born to a Hokkaido apple tree pruner, Nagayama had grown up in poverty. His father's drinking and gambling had burdened the family with heavy debts and to provide the barest necessities his mother peddled fish on the street. When his eldest brother fathered an illegitimate child it fell to Nagayama's mother to raise it. At about this time his eldest sister developed severe psychological problems which placed additional strain on family relationships. To supplement the mother's meagre earnings the children did any work available including delivering newspapers and collecting scrap iron. Eventually the father left the family to live on the streets. He died soon after. These were the circumstances surrounding Nagayama as he advanced through junior high school. The junior high school administration took pity on him and allowed him to graduate even though his attendance and performance records were below the acceptable level.

In the mid-sixties, when Nagayama graduated, junior high school graduates were in high demand as blue collar workers. Nagayama and twenty of his fellow graduates went to Tokyo looking for work. He accepted a placement with the Nishimura Fruit Parlour which was in need of one new employee only. By choosing to work apart from his friends Nagayama wanted to sever his ties with the past and begin anew. He had brought his textbooks to Tokyo and planned to further his education through evening high school courses. However, because of the circumstances under which he had been allowed to graduate, his junior high school refused to issue the requisite transcript thus effectively barring Nagayama from further education. When Nagayama realized he was powerless to change the situation he hid his profound disappointment beneath a 'couldn't care less' attitude.

Before long fighting and rabble-rousing resulted in his dismissal from the fruit parlour. Still seeking another chance, this time in a foreign environment, he went to Yokohama harbour and stowed away on a Danish ship. He was discovered and deported from Hong Kong. After this Nagayama's career was varied. It included apprenticeship in automotive repair, rice and milk delivery, waiting on tables in a teahouse, and eventually entry into the world of crime. In 1966 he was arrested after stealing money at the American naval base at Yokosuka. Later that same year he was apprehended after attempting to rob a store in Utsunomiya city. As a juvenile offender the penalty even after a second offence was not severe. In October 1968, he returned to Yokosuka and stole the revolver with which he would commit four murders.

Two movies, *Hadaka no Jukyusai* (Naked at Nineteen) and *Ryakusho*

Renzoku Shasatsuma (So-called Serial Killer), based on his life were made after the publication of his books. The books and the movies drew attention to the societal conditions which contributed to the development of Nagayama's criminal behaviour. He designated the families of his victims the recipients of the royalties. (The sequence of events in Nagayama's initial court case is described by Toshio Yanagi, Crown Prosecutor, Criminal Affairs Bureau, Ministry of Justice in 'Nagayama Saikosai Hanketsu – Sono Keii to Gaiyo' (Supreme Court Decision on Nagayama's Case – Summary of its Evolution), *Horitsu no Hiroba* (Law Forum), 1983, vol. 36, no. 10, pp. 4–11.

11 W.A. Bonger, *Criminality and Economic Conditions*, H.P. Horton (trans.), Boston, Little, Brown, and Company, 1916, p. 436.

12 In his discussion of the communicability of innovations Rogers cites a study by E. Hruschka (unpublished data, Stuttgart-Hohenheim, Germany, Institute für Landwirtschaftliche Beratung an der Landwirtschaftlichen Hochschule, 1961). Hruschka found that 'observability' was a key factor in the rapidity of diffusion of new ideas. For example, when new techniques were introduced by demonstration farmers in Germany, more villagers in the surrounding area were aware of a new haymaking technique than of new ways of keeping farm records.

13 *Nihon Keibi Hosho Company Song*
> Rain at night; a blizzard in the morning,
> We guard with courage.
> With effort and diligence and service in our hearts,
> We protect this day.
> Our shining company; our shining company,
> Oh! Nihon Keibi Hosho!
>
> We are proud of our corporate symbol: key and owl,
> We proceed with determination.
> Forever engraved in our hearts,
> Sincerity, responsibility, alertness.
> Our prospering company; our prospering company,
> Oh! Nihon Keibi Hosho!
>
> Go forth today, prepare for tomorrow,
> Conquering hardships we smile.
> Performance and hope we are filled with joy,
> Shoulder to shoulder we sing together.
> Cherish our company; cherish our company,
> Oh! Nihon Keibi Hosho!

6 INTERNAL AND SOCIETAL ADJUSTMENTS

1 R.M. Kanter, *The Change Masters: Corporate Entrepreneurs at Work*, London, Allen & Unwin, 1983.

2 Keisatsucho Keisatsushi Hensan Iinkai, *Sengo Keisatsu-shi* (Postwar Police History), Tokyo, Keisatsu Kyokai, 1977, p. 50.

3 J.E. Tropman and G. Morningstar, *Entrepreneurial Systems for the 1990s:*

Their Creation, Structure, and Management, New York, Quorum Books, 1989, p. 43.

4 Japan National Police Agency, *1976 Keisatsu Hakusho* (1976 White Paper on Police Services), Tokyo, Ministry of Finance, 1976, p. 49.
5 These events were reported in *Asahi Shimbun*, 12–20 January 1971.
6 These events were reported in *Asahi Shimbun*, 24–7 February 1971.
7 Japan National Police Agency, *1983 Keisatsu Hakusho* (1983 White Paper on Police Services), Tokyo, Ministry of Finance, 1983, pp. 27–9; There were 136 guerrilla incidents between 1984–8.
8 By 1989, of the 2,165 officially recognized sufferers of Minamata disease, 800 had died. Another 5,200 cases were under consideration for 'official recognition'.
9 For a discussion of *sokaiya* see K. van Wolferen, *The Enigma of Japanese Power*, New York, Alfred A. Knopf, 1989, pp. 105–7.
10 These events were reported in *Asahi Shimbun*, 26–9 May 1971.
11 These figures were extracted from the National Police Agency surveys of 1971 and 1970 and reported in *Asahi Shimbun*, 28 May 1971.
12 *Sengo Keisatsu-shi*, op. cit., p. 836.
13 *1976 Keisatsu Hakusho*, p. 50.
14 *Sengo Keisatsu-shi*, op. cit., p. 1162.
15 *Keibi Hosho Shimbun*, 19 and 26 June 1980.
16 *1976 Keisatsu Hakusho*, p. 50.
17 Japan National Police Agency, *1987 Keisatsu Hakusho* (1987 White Paper on Police Services), Tokyo, Ministry of Finance, 1987, p. 147.

7 CULTIVATION OF HUMAN RESOURCES

1 For example, on one occasion, talking to middle managers, Iida elaborated on a list of fifteen items he felt were crucial to their conduct.

i   Managers must understand the primary reason for the company's existence and what it is striving to become in the future. This will be the standard against which decisions are made.
ii   Managers must recognize the reasons for the formulation of corporate rules. They must ensure understanding of these rules on the part of subordinates and, based on this understanding, see that the rules are enforced.
iii To run the organization rules are necessary. In addition to understanding the fundamental and secondary rules managers must develop an understanding of the relative importance and sequence of these rules and conduct business accordingly.
iv   There are always a variety of solutions to problems and ways in which tasks can be performed. It is always important to seek alternatives and select from among more than two.
v   When a manager is confronted by a difficult problem his ability as a manager will be utilized and his true quality manifested.
vi   Believe in your own potential and that of the company showing courage to challenge this potential.
vii Managers should make themselves clearly understood.

viii Towards the corporation, clients, colleagues, and subordinates, of primary importance is honesty and sincerity.

ix A manager is expected to be a respected member of the corporation but it is important to realize that this is dependent upon his being a respected member of society. He must act accordingly.

x The morale and behaviour of subordinates reflect the morale and behaviour of managers. When harmony is absent or morale low the manager must examine his spirit and behaviour.

xi In order to complete a task, in addition to employee theories and reason, it must be understood that if human emotions are not taken into consideration success will be impossible.

xii Managers must understand that accomplishments belong to their subordinates and the mistakes of subordinates are those of the managers.

xiii Managers should reflect on their feelings at times when they are treated unjustly and should try to be fair to all.

xiv Leadership is not shown by words; it is shown by practice.

xv Growth of subordinates is proof of a manager's ability. A manager should be proud of this growth.

2 G. Orwell, *Nineteen Eighty-Four*, London, Secker and Warburg, 1949.

3 Y. Aonuma, *Nihon no Keiei-so* (Senior Executives in Japan), Tokyo, Nihon Keizai Shimbunsha, 1965, p. 161.

8 INNOVATIVE VENTURES

1 Z. Katagata recently published *Shakai Shisutemu Sangyo: SECOM Kaicho Iida Makoto no Senryaku to Hasso* (Social System Industry: Strategy and Inspiration of Makoto Iida, Chairman of SECOM), Tokyo, Kodansha, 1989. Other books which have as their focus Makoto Iida and SECOM's venture into the information industry, all written by journalists, include Y. Sakaguchi, *SECOM: Iida Makoto no Jyoho Senryaku* (SECOM: Makoto Iida's Information Strategy), Tokyo, Paru Shuppan, 1986; A. Iwabuchi, *Shakai Shisutemu SECOM no Zenbo* (Total picture of SECOM as a Social System Industry), Tokyo, Yamate Shobo, 1986; Y. Akiba, *SECOM Mirai Jyoho Bijinesu: Network Jidai no Kigyo Senryaku* (SECOM, a Future Information Business: Corporate Strategy of Network Age), Tokyo, TBS Britannica, 1987; A. Iwabuchi, *No.1 Jyoho Network no Himitsu: 21 Seki o Mezashita SECOM no Mirai Senryaku no Zenyo* (Secret of the No.1 Information Network: Total Picture of the Future Strategy of SECOM as the 21st Century Approaches), Tokyo, Diamond-sha, 1989.

2 Kasai Yobo Gyosei Sanjunenshi Hensan Iinkai, *Kasai Yobo Gyosei Sanjunenshi* (Fire Prevention Administrative History over 30 Years), Tokyo, Tokyo Shobocho, 1978, p. 544.

3 Hyakunen Kinengyoji Suishin Iinkai, *Tokyo no Shobo Hyakunen no Ayumi* (The Evolution of Fire Prevention in Tokyo over the last 100 years), Tokyo, Tokyo Shobocho, 1978, p. 136.

4 *Kasai Yobo Gyosei Sanjunenshi*, op. cit., p. 89.

5 'Kyodo-jutaku ni okeru Kasai Kiken no Kaimei to Jinmei Anzen Taisaku'

(The Identification of Fire Hazards in Residential Buildings and Steps Towards Fire Prevention), *Tokyo Shobo*, May 1987, p. 16.

6 For a discussion of future business development in terms of products and markets as this relates to the process of vision definition see B.B. Tregoe, J.W. Zimmerman, R.A. Smith and P.M. Tobia, *Vision in Action: Putting a Winning Strategy to Work*, New York, Simon and Schuster, 1989, pp. 36–69.

7 A. Ueda, *Nihon-jin to Sumai* (Japanese and Their Houses), Tokyo, Iwanami Shoten, 1974, pp. 208–9.

8 Ueda points out that the spatial pattern of burglary in Kyoto has closely followed the outward spread of the nuclear family from the city centre to the urban fringe.

9 The Japanese government is exploring nuclear power, imported coal, LNG, liquified coal gas, solar heat, solar light, geothermal heat, and alcohol as potential alternative energy sources. The Organization for General Development of New Energy has been established to carry out research and development in these fields. Small- and medium-sized hydroelectric systems and systems using biomass, wind power, wave power, and ocean temperature differentials are also under investigation as alternative energy sources.

10 For a discussion of the competitiveness of coal vs uranium as a power generating fuel in selected western countries, see G. Lermer, *Atomic Energy of Canada Limited: The Crown Corporation as Strategist in an Entrepreneurial, Global-Scale Industry*, Ottawa, Minister of Supply and Services Canada, 1987, pp. 47–50.

11 American nuclear powered vessels were also the target of mass protests. It was not until 1964, after years of negotiations to ensure their absolute safety, that nuclear powered submarines were allowed into naval bases in Japan, initially in the face of massive demonstrations. With feelings over the Vietnam War running high, the visit of an American nuclear powered aircraft carrier in 1968 occasioned even louder protests. Japan's own experiment with a nuclear powered vessel, the Mutsu, ended in severe embarrassment for the government when no Japanese port community would agree to serve as a home port for the ill-fated vessel.

12 K. van Wolferen, *The Enigma of Japanese Power*, New York, Alfred A. Knopf, 1989, p. 194.

13 S. Ito (ed.), *Toshi to Hanzai* (Cities and Crime), Tokyo, Toyo Keizai Shimpo-sha, 1982, pp. 101–2. Japan Office of the Prime Minister, 'Mijikana Hanzai ni Kansuru Yoron Chosa' (Public Opinion Survey with regard to Assault and Burglary), reported in *Bohan Setsubi* (January 1987), pp. 97–123.

14 *Daily Tohoku* (Aomori), 22 October 1989.

9 VERTICAL INTEGRATION

1 M. Moritani, *Japanese Technology*, Tokyo, Simul Press, 1982 pp. 7–8.

2 Norman Macrae, 'The Coming Entrepreneurial Revolution: A Survey', *The Economist*, 25 Dec. 1976, p. 42.

10 INTERNATIONALIZATION

1 F.K. Reilly, *Investment Analysis and Portfolio Management*, Chicago, Dryden, 1985, pp. 741–4.
2 This information was provided by Motoko Katakura, research scholar at the National Museum of Ethnology, Osaka.
3 *Republic of China Almanac 1988*, Tokyo, Taiwan Research Institute, 1988, p. 26.
4 There are some 200 mountains over 3,000 metres high in Taiwan.
5 *Republic of China Almanac 1988*, op. cit., p. 1246.
6 'Gang Crime in Taiwan', *China Post*, Taipei quoted in *World Press Review*, December 1990, p. 58.
7 *Republic of China Almanac 1988*, op. cit., p. 1256.
8 For more discussion, see M. Matsumoto, *The Unspoken Way: Haragei, Silence in Japanese Business and Society*, Tokyo, Kodansha International, 1988.
9 M. Davis, *Prevent Burglary!: An Aggressive Approach to Total Home Security*, Englewood Cliffs, Prentice-Hall, 1986, p. 2.
10 R. Arnold, ' "Laverne" of TV Trips Up Two Robbers in Her Home', *Los Angeles Times*, 15 March 1984.
11 Tokyo Shobocho, *Tokyo Shobo Hyakunen no Ayumi* (100 Years of the Tokyo Fire Defence Agency), Tokyo, Tokyo Shobocho, 1980, pp. 500–6.
12 *Nihon Keizai Shimbun*, 14 August 1990.
13 ibid.
14 Sogo Keibi Hosho, SECOM's number one competitor, in conjunction with thirteen other companies, moved into the area of emergency taxi service and established Home Net Corporation in August 1989. Clients are provided with a push button control with which to summon a taxi when needed. Japan's 256,000 taxis carried 3.3 billion people in 1989. This is far below capacity and concern about insufficient business is widespread, especially in rural areas where the number of private cars is on the increase. By the end of 1989 twenty taxi companies throughout Japan had entered into franchise agreements with Sogo Keibi Hosho in order to receive marketing and operational know-how in order to handle emergency situations.
15 E. Clifford, 'Treating Patients at Home a Growth Industry', *Globe and Mail*, 6 October 1990.
16 Advancement of Japanese corporations into the American movie-making industry is proceeding rapidly. Not only is the movie industry itself profitable but the resulting video production directly affects the sale of video cassette recorders. Increased software access is the primary reason for Nihon Victor's sizable investment in movies produced by Lawrence Gordon. Many American film investment corporations have sprung up in Japan. SECOM together with nine other companies including Olix, Nippon Securities, and Terada Warehouses, in 1989 created the American Movie Making Investment Fund.

11 RESEARCH – FOUNDATION FOR GROWTH

1 T. Kono, 'Factors Affecting the Creativity of Organizations – An Approach

from the Analysis of New Product Development', in K. Urabe, J. Child and T. Kagono (eds), *Innovation and Management: International Comparisons*, Berlin, Walter de Gruyter, 1988, p. 112.

2 For a discussion of the nature of NTT Tsukens's connections to the academic and industrial sector with regard to personnel transfer and joint research, see 'Kensho NTT Tsuken' (Verifying the nature of the problems of NTT Tsuken), *Nikkei Communications*, No. 78, 12 March 1990, pp. 77–124.

3 D. McMann, 'Good Home Security', *Westworld*, July/August 1990, vol. 16, no. 3, pp. 6–9.

4 'Setting the Pace in R&D: 1989 White Paper on Science and Technology, Science and Technology Agency', 'Japanscene', *Focus Japan*, May 1990, vol. 17 no. 5.

5 G. Lermer, *Atomic Energy of Canada Limited: The Crown Corporation as Strategist in an Entrepreneurial, Global-Scale Industry*, Ottawa, Minister of Supply and Services Canada, 1987, p. 22.

6 'Setting the Pace in R&D: 1989 White Paper on Science and Technology, Science and Technology Agency', op. cit.

12 MAKING HISTORY

1 G.L.S. Shackle, foreword to R.F. Hébert and A.N. Link, *The Entrepreneur: Mainstream Views and Radical Critiques*, New York, Praeger, 1982, p. viii.

2 A. Iwabuchi, *Shakai Shisutemu SECOM no Zenbo* (SECOM as a Social Systems Industry: Total Picture), Tokyo, Yamate Shobo, 1986, pp. 181–2.

13 CONCLUSION

1 P.H. Wilken, *Entrepreneurship: A Comparative and Historical Study*, Norwood, New Jersey, Ablex, 1979, pp. 67–8.

2 W.P. Glade, 'Approaches to a Theory of Entrepreneurial Formation', *Explorations of Entrepreneurial History*, 1967, series 2, vol. 4, pp. 245–59; D.C. McClelland and D.G. Winter, *Motivating Economic Achievement*, New York, Free Press, 1971. Cited in P.H. Wilken, op. cit., p. 68.

3 W.E. Peacock, *Corporate Combat*, New York, Facts on File Publications, 1984, p. xi.

4 P.H. Wilken, op. cit., pp. 160–90.

5 K. Kamioka, *Japanese Business Pioneers*, Union City, CA, Heian International, 1988, pp. 60–75.

6 ibid., pp. 105–6.

7 H. Tanaka, *Personality in Industry: The Human Side of a Japanese Enterprise*, London, Pinter, 1988.

8 K. Ohmae, *The Mind of the Strategist: The Art of Japanese Business*, New York, McGraw-Hill, 1982, pp. 194–5.

9 T. Sakiya, *Honda Motor: The Men, the Management, the Machines*, Tokyo, Kodansha International, 1982.

10 A. Morita with E.M. Reingold and M. Shimomura, *Made in Japan: Akio Morita and Sony*, New York, E.P. Dutton, 1986.

11 L. Smith, 'Japan's Autocratic Managers', *Fortune*, 7 January 1985, pp. 56–65.

APPENDIX: A SOCIAL SCIENTIST'S SUMMARY OF ENTREPRENEURSHIP

1 R.F. Hébert and A.N. Link, *The Entrepreneur: Mainstream Views and Radical Critiques*, New York, Praeger, 1982, p. ix.
2 T.C. Cochran, 'Entrepreneurship', *International Encyclopedia of the Social Sciences*, vol. 5, New York, Macmillan, 1968, pp. 87–90. For a discussion of the major contributions of social scientists working in this area, see R.C. Cauthorn, *Contributions to a Theory of Entrepreneurship*, New York, Garland, 1989.
3 J.A. Schumpeter, *The Theory of Economic Development: An Inquiry Into Profits, Capital, Credit, Interest and the Business Cycle*, Cambridge, Harvard University Press, 1934; *Business Cycles: A Theoretical, Historical, and Statistical Analysis of the Capitalist Process*, 2 vols., New York, McGraw-Hill, 1939; *History of Economic Analysis*, New York, Oxford University Press, 1954.
4 T. Parsons, *The Structure of Social Action*, New York, McGraw, 1937; T. Parsons and N.J. Smelser, *Economy and Society: A Study in the Integration of Economic and Social Theory*, Glencoe, Free Press, 1956.
5 H.G. Barnett, *Innovation: The Basis of Cultural Change*, New York, McGraw-Hill, 1953, p. 1.
6 A.H. Cole, *Business Enterprise in its Social Setting*, Cambridge, Harvard University Press, 1959, p. 7.
7 R.C. Cauthorn, op. cit., p. 125.
8 P.S. Christensen, *Strategy, Opportunity Identification, and Entrepreneurship: A Study of the Entrepreneurial Opportunity Identification Process*, Aarhus, Institute of Management, University of Aarhus, 1989, pp. 4–5.
9 H. Mintzberg, 'Strategy-Making in Three Modes', *California Management Review*, Winter 1973; H. Mintzberg and J.A. Waters, 'Tracking Strategy in an Entrepreneurial Firm', *Academy of Management Journal*, 1982, vol. 25, no. 3, pp. 465–99; M. Porter, *Competitive Strategies*, New York, Free Press, 1980; P. Drucker, *Innovation and Entrepreneurship*, New York, Harper and Row, 1985; A.D. Chandler, Jr, *Scale and Scope: The Dynamics of Industrial Capitalism*, Cambridge, Harvard University Press, 1990.

Index

Abo, F. 160–2, 146
academic conferences, participation
    in 208, 211
acquisitions and mergers: by
    Japanese companies 187–8; by
    SECOM 153–5, 187
Ademco sensors 143
administrative behaviour 3
advertising 89
age, relative 88
agricultural technology 229–30
Aiba, H. 45
airport, Narita 100–2
alarms and alarm systems 68–76,
    132, 137–8, 204
alertness 45
alienation 113
amakudari 87
ambulance services 184–7
American Electronics Association
    209
Aoyama Auditing Corporation 224
arrogance 23
artificial intelligence research 210
Asado, Y. 51–2, 62–3
audit networks 214
audits: external 53; system 224
Australia, SECOM activities in 155–
    6
autocracy 238
automated banking services 126–7

babysitting services 226
bankruptcy 106
banks: computerisation 126–7;
    lending by 42, 162; robberies 74–

5; security systems 73, 126–7; see
    also financial institutions
Bernard, C. I. 243
Barnett, H. G. 243
bedridden people, alarms for 137–8
behaviour, correct 17–18, 20
betting 57
bicycle transport 73–4
bio-technology 229–30
biometric entry control 134
black market 19–20
Bohan Kyokai 39, 104, 105
bond issues 153
borrowing for business launch 41–3,
    126
boryokudan 19, 103
Bottomly, C. 114
bowling alleys 29–30
Brambles Security Group 155–6
broadcast dramatisation of security
    industry 63–4, 76
Buddhism 40
budgetting 50, 211
building services 139–42
burglary 133, 136
business ethics 20, 69, 79, 234
business plans 41–2
buying-on-demand 50

cable TV 228
Calgary 205
CAPTAINS 215, 216, 217
Care Visions Inc. 188–9
cash flow 48
CATV 178, 215, 228
CD Security Pack 127

Central Training Institute, SECOM
114–15
chairman of NKH/SECOM 121
change 23, 95, 97, 236–7
Chiba, S. 70–1
chief executive officer: NKH/
SECOM 121; SECOMERICA
181–4
children, services for 188–9
chimney fires 131
Chisso Corporation 102–3
Chiyoda Corporation 156
chlorofluorocarbons (CFCs) 203
Chokwatana, B. 193–4
Chung Hsing Security
Communication Co. 159–68
Chuo Keiba Horse Racing
Association 57
coalitions 88
code of conduct for security
industry 69
Cole, A. H. 243
commitment 113, 239
communication: between researchers
212; via electromagnetic waves
204–5; with guards 113;
interpersonal 172, 182–3;
networks 125–6, 214–25; within
NKH/SECOM 89, 211, 218
community antenna TV (CATV)
178, 215, 228
company songs 84–5, 149–51
competition in security industry 64–
7, 106
competitive behaviour 3
Computer Service Kaisha (CSK)
221
computerisation: of banks 126–7; of
manufacture 144; of NKH/
SECOM 74, 146; of security
systems 91, 124–6, 146
computerised information services
214–25
computers, security of 224–5
concerts 32
conferences, participation in 208,
211
conformity of Japanese 29, 49
Confucianism 10
consensus 238

consequences of actions 231
construction industry 94
contract manufacturing 144
control of security industry 98–107
convertible bond issues 153
corporate culture of NKH/SECOM
69, 87, 88, 112, 148–9, 184, 190–1,
208, 234
corporate labour unions 59
corporate logos 44, 66, 196
correct behaviour 17–18, 20
costs: of information services 217;
of labour 94; of training 94
courage 212
creativity 4, 5, 31, 88
credit 47–8; Lida's dislike of 24; *see
also* loans; prepayment crime by
employees 77–80
crime prevention associations 39,
104, 105
crime rate: Japan 39, 40, 136;
Taiwan 167–8; Thailand 192;
USA 176
criminals, capture of 81–3
crisis responses 83
crowd control 32
cryptography 225
cultivation technology 229
customers, importance of 184, 188,
191

Dai-ni Den Den 215, 216
data collection and analysis 58
database services 214–25
debt of Korea SECOM 173
decision making by consensus 238
decisiveness 212
Denki Tsushin Kenkyujo (Tsuken)
200–1
determination 78
disasters 124
diversity of employees 49, 89–90
dormitories 120
doryo 17
dramatisation of security industry
63–4, 76
drop-in child care 226–7
Drumm, M. 212
Dubos, R. 12
duty 46

earthquakes 145–6
eavesdropping 229
economies of scale/scope 75
education: of Iida 12–18; of NKH/
    SECOM employees 118–21, 190;
    social 17–18; *see also* training
eel farms 128
elderly people, services for 137–8,
    185
electronic alarm systems 68–76
electronics industry information
    service 218–19
elitism 208
emergency alarms 137–8
emergency medical services 185–7
Emergency System (ES) Project,
    SECOM 204
employees: external environment of
    89–90; morale 68–9, 107;
    orientation 45; personality 49;
    routine work 73; screening of 79;
    selection *see* recruitment; self-
    actualization 111; theft by 77–80,
    107
energy conservation systems 141
Engineering and Development
    Centre, SECOM 199
English Language 63, 119, 236
entrepreneurship 2–4, 242–3
entry control 132–4, 136
environmental control technology
    229
ES (Emergency System) Project,
    SECOM 204
espionage 76–7
ethical business 20, 69, 79, 234
etiquette 17–18
exchange of knowledge 123–4
exhibitions 93
expenses of business launch 41–3
external audits 53
external environment of employees
    89–90
Ezawa, Y. 119–20, 192–7

facsimile communication (fax) 229
false alarms 132, 204
family business of Iida 18–21, 35
family life of Iida 231–2
fate 40

FEAL-8 225
festivals 32
financial institutions: security
    systems for 172, 175; *see also*
    banks
financial projections 41–2
financing of business launch 41–3
fingerprint controlled entry 134
Fire Alarm Full Automatic 130–1
fire alarm systems 129–32
Fire Defence Agency (Japan) 129,
    131, 186
fire extinguishing systems 129–32,
    203
fire prevention 129–32, 136
fire security 128–32, 168
firearms 40–1, 77
fish farms 128
fitness clubs 226–7
flame sensors 131
fluorine 203
football, Iida's interest in 16–18, 28
foreign *see* overseas
friends of Iida 26–9
FS (Future Structure/Full Support)
    Centre, SECOM 221
Fujisawa, T. 238
funding of research 211–12
future, prediction of 37, 71

Gakushuin University 15–18, 28
gangs 19, 101
gas sensors 136
giri 46
Gloria Travel Agency 227–8
go-betweens 46
Goto, N. 63, 158
government funding of research 211
growth of NKH/SECOM 67, 68,
    86, 95, 108–11, 240
The Guardian (*Za Gardoman*) 64,
    76
Guilford-Martin Personnel
    Inventory 45, 49
guns 40–1, 77

halon 130, 203
handicapped people, services to 187
haragei 172
Hashimoto, S. 115, 206–13
Hattori, K. 149

Hayashi, T. 52–3
health services 184–9, 225–7, 230–1
heat sensors 130
helicopter transport services 228
High Plant, SECOM 229
Hirata, S. 119
Hitachi Electron 70
Hitotsubashi Business School 81–3, 89
HMSS Inc. 187–8
Hochiki 131
holiday resorts 117
home health care 187
home security systems 136–8, 228
homogeneity of Japanese 29, 49
Honda, S. 238
honesty and integrity 11, 20, 234
honour 46, 79
horse racing 57
horticultural technology 229–30
hospital ambulance services 185
Hospital Corporation of America (HCA) 185
hotel fires 131
hotel security 62–3
household security 133, 136–9
Hsiao Cheng-chih 160–2
Human Development Centre, SECOM 115–17
humility 23, 24
Huntington National Bank 126
hydroponics 229–30

IBM 221
Ibuka, M. 238
Ieyasu Tokugawa 8–9
Iida, Atsushi (brother of MI) 11, 36
Iida, Eikichi (grandfather of MI) 10, 234
Iida, Fusa (mother of MI) 8, 23–4, 138
Iida, Goichi (son of MI) 25
Iida, Hiroshi (brother of MI) 11, 20, 24, 26, 35–6
Iida, Mami (daughter of MI) 60, 231–2
Iida, Misao (wife of MI) 25
Iida, Monjiro (father of MI) 8, 11, 12, 19–21, 22–3, 24, 26, 78, 231, 234

Iida, Susumu (brother of MI) 11, 20, 23, 24, 26, 35
Iida, Tamotsu (brother of MI) 11, 20, 24, 26, 34–5
Ikeda, H. 48
illegal trading 19–20, 234
image of security industry 84
image processing research 210
Inamori, K. 216
independence of Iida family 26, 28
information services, computerised 214–25
information systems in buildings 141–2
infra-red controlled entry 134
infusion therapy services 187, 188
initiative 49
innovation 4, 29, 41, 46, 50, 123–42, 236–7; planning of 71; technological 68–85, 95, 187, 214
integrated networks 140
integration: of alarm systems 75; of people and technology 72
integrity and honesty 11, 20, 234
intelligent buildings 141–2
Intelligent Systems (IS) Laboratory, SECOM 200, 205–13
intermediaries 46
international expansion of SECOM 152–97
international security federation 42–3, 72, 84, 103, 114
interpersonal communication 172, 182–3
interviews in staff selection 49, 51
intravenous therapy services 187, 188
Inumaru, T. 61–2
investment in innovation 50
IS (Intelligent Systems) Laboratory, SECOM 200, 205–13
Isetan 79
Ishihara, S. 15
Ishihara, T. 120
Ito, F. 56

Japan Association of Female Executives (JAFE) 226
Japan Nuclear Security System (JNSS) 135

Japanese language training 210
Jastic Kansai 51
jitsugyoka 10
judgement 214
Junet 212
just-in-time production 237

Kajiwara, T. 156
kanban 237
Kangaroo Kids 189
Katagata, Z. 123
Kato, Z. 89
Kawasaki 137
Kaye, M. 18, 90, 179–85, 186, 189–91
keibi hosho 43
Keizai Doyukai 120
keys and locks 132–4
Kimura, K. 14, 86–7, 220
Kimura, S. 90–1
Kirishima, H. 119
kitchen chimney fires 131
knowledge, exchange of 123–4
Kofuku-ji temple 92
kohai 17
Kojima, M. 148
Komatsuzaki, T. 225, 227
Komine, K., 87–8
Korea, SECOM activities in 169–76
Korea SECOM 171–6
Korea Security Corporation 171
Koushan Enterprises 159
Kyocera 216
Kyodo VAN 221
Kyokusa Boryoku Shudan 76, 101

labour costs 94
labour negotiations 59–60
labour unions 58–60
language: English 63, 119, 236; Japanese 210
leadership 79
leased private circuits, 124, 125–6
leasing of equipment 162
Lee Byung-chull 169, 174
Lee Dong-woo 175–6
Lee Kun-hee 174–5
legal control of security industry 98–107
licensing of security companies 104
LifeFleet 185

Ligue Internationale des Sociétés de Surveillance 42–3, 72, 84, 103, 114
Lin Roscow 160–4, 166–8
Lin Teng 158–63
liquidity 48
liquified petroleum gas (LPG) monitoring 221
loans for business launch 41–3, 162
locks and keys 132–4
logical thinking 37
logos, corporate 44, 66
Lowebigham and Thomson 53

Macrae, N. 143–4
magnetic card controlled locks 134
mail order 30
managers: recruitment 86–90, 153, 239; training 22
manufacture of security systems 143–5, 163, 177
market segmentation 219
marriage of Iida 25, 60–1
Massachusetts Institute of Technology (MIT) 209
Masuda, I. 211
Matsuo, Y. 118
Matsushita, K. 237
Matsuyuki, M. 176
Matsuzaki, N. 156, 157
media networks 214
medical sensors 203
medical services 184–9, 225–7, 230–1
Meiji era 9–10, 234
mergers *see* acquisitions
microelectronics in NKH 143
microwave transmission of sensor signals 199–200
militarism 40
Ministry of International Trade and Industry (MITI) 216, 219, 224
Mitsubishi Bank 73
Miyagi Network 228
Miyauchi, S. 201–2, 207
mobile depots 96
money as incentive 34
morale of employees 68–9, 107
Morita, A. 238
motivating people 90
motivation 113

Murai, J. 64–6
musical performances 32
Mutsu 138
My Alarm 136–8
My Care 230–1
My Doctor 137–8, 204
My Fit 226–7
My Poppins 226–7

Nagashima, Y. 24, 180
Nagayama, N. 82–3
Nakaminato city hall 98–100
Nakamura, N. 226
Nakane, C. 17
Nakaya, T. 81–2, 83, 91
name change to SECOM 97–8, 146–7
Narikawa, N. 144–5
Narita airport 100–2
National Federation of Crime Prevention *see* Zenkoku Bohan Kyokai Rengo-kai
navy, Iida's ambitions toward 13–14
networks: communication 91, 125–6, 214; community antenna TV (CATV) 178, 215, 228; integrated 140; social 214
new activities, challenge of 15
new business startups 36–7
new ideas 95, 237
new materials technology 229
newness, perception of 30–1
Nihon Keibi Hosho Kabushiki Kaisha 7, 43, 97–8, 146
Nihon Keibi Hosho Printing Centre 149
Nikkei Densetsu Corporation 91
Nikkei McGraw-Hill database 219
Nippon Computer Security Corporation (NICOSE) 224–5, 229
Nippon Denshin Denwa Kosha *see* Nippon Telegraph and Telephone
Nippon Norin Helicopter Corporation 228
Nippon Sekiyu Gasu Corporation 221
Nippon Telegraph and Telephone (NTT) 70, 87, 200–1, 206, 208, 215, 216, 217, 224, 225

Nishimura, Y. 169–70, 172–3
Nohmi Bosai 131
NTT *see* Nippon Telegraph and Telephone
nuclear security 134–5
nursing services 188–9

obsolescence 95
ocean, Iida's feelings towards 14–15
Office of Technology Investigation, SECOM 204
oil refineries 156–7
Okabe, J. 54
Okada, K. 73–4
Oki Electric Industry 144–5
old people, services for 137–8, 185
Olympic Games, Tokyo 53–7
on-line home security systems 136–8
Onodera, H. 165
open thinking 68
optical entry systems 134
orientation of employees 45
originality 31
Osada, S. 24–5, 39–40, 74, 89
Osaka 60–1
overseas expansion of SECOM 152–97
overseas influences on Iida 236
overseas investment: by Japanese companies 153–5, 187–8; in security industry in Japan 65, 77
overseas research staff 209
overseas scholarships 118–21
overseas students in Japan 120–1
owl logo 44

paramedics 186
Parsons, T. 242
patient care 188–9
patrol cars 104, 196
pay 58–9
paying for services 217–18
payment-in-advance 29, 33, 46, 47–8, 55, 234, 239
peace of mind 185, 225
pediatric services 188–9
people and technology: integration of 72
perception of newness 30–1
periodic security checks 94–5
persistence 239

personal alarms 137–8
personality of employees 49
personality of Iida 90
petrochemical plants 156–7
planning of innovation 71
Pochen, K. 209
police in Japan 31–2, 39, 56, 57,
    103, 104
predetermined destiny 40
predicting the future 37,71
prepayment 29, 33, 46, 47–8, 55,
    234, 239
president: NKH/SECOM 122;
    SECOMERICA 181–4
press relations 77
price of services 217–18
Price Waterhouse 53, 224
privacy 104, 134, 210, 229
private circuits 124, 125–6
private profit 10
product development 198–213
profit, private 10
proper behaviouir 17–18, 20
public image of security industry 84
public relations 74, 89
push-button locks 134
Pythagoras safe 127

qualifcations of security guards 104
quality control in manufacuture
    163, 203, 204–5
questioning 68

radio communication with guards
    113
radio transmission of sensor signals
    199
reading habits of Iida 38
recruitment 45, 49, 94; managers
    86–90, 153, 239; research staff
    207; South Korea 171–2; Taiwan
    164; Thailand 195
regulation of security industry 98–
    107, 168
relative age 88
religious festivals 32
remote alarm systems 68–76
research and development 124, 198–
    213
resistance to change 97
resources 5

response time 125
responsibility 45
risks and risk management 141, 214
robberies 133, 136; of banks 74–5;
    by employees 77–80, 107
robots 210–11, 229
routine work 73
Ryad Oil Refinery 156–7

safes 127
safety of public 33, 39, 124
Saito, S. 80–1
Sakaguchi, T. 113
sake industry 10, 18–21, 22–4, 35–6
salaries 58–9
Samsung Company 169–72, 174
Sasaki, M. 82, 83, 91
satellite TV 228
Satha Pathana Group 193–4
Sato, H. 84
Satori, A. 198
Sato, Y. 112–13
Saudi Arabia, SECOM activities in
    156–8
scandals affecting security industry
    98–103
scholarships 118–21
school security systems 175
Schumpeter, J. A. 242–3
Science and Technology
    Foundation, SECOM 124
scramblers 210, 229
screening of employees 79
SD (System Design and System
    Development) Centre, SECOM
    145–6
SECOM 5–7, 43, 96–8, 146–51, 236
SECOM-24 227
SECOM Central Training Institute
    114–15
SECOM Engineering and
    Development Centre 199
SECOM ES (Emergency System)
    Project 204
SECOM FA2 130–1
SECOM FS (Future Structure/Full
    Support) Centre 221
SECOM High Plant 229
SECOM Human Development
    Centre 115–17

SECOM Industry 144–5
SECOM IS (Intelligent Systems) Laboratory 200, 205–13
SECOM Maintenance 91, 139
SECOM Office of Technology Investigation 204
SECOM Printing Centre 149
SECOM Science and Technology Foundation 124
SECOM SD (System Design and System Development) Centre 145–6
SECOM Service 91
SECOM Systems Company 139
SECOM Technical Centre 198–205, 207
SECOM Totax-T 141
SECOMERICA 184–91
SECOMERICA University 190
Secomnet 220–1
Securitas sensors 143
Security (journal) 123
security of computers 224–5
Security Guard Industry Act 104–5, 107
security industry: broadcast dramatisation of 63–4, 76; code of conduct 69; competition in 64–7; international cooperation 42–3, 72, 84, 103, 114; overseas investment in 65; public image 84; regulation of 98–107
security lock system 134
Security Patrol Company 43
Security Patrol (SP) Alarm System 71–2, 84, 89, 91, 95–6, 124, 126, 140, 225
security robots 210–11
Security World Corporation 123
Seki, G. 58–9
selecltion *see* recruitment
self-actualization 111
self-confidence 23
self-discipline 231
self-motivation 191
sempai 17, 88
Senno, K. 173–4
sensitivity 212
sensors: for building control 140, 203; fire 136; flame 131; gas 136;

heat 130; malfunctions 204–5; medical 203; research on 200, 202–3; smoke 130; sources of 143; transmission of signals from 199–200
service ethos 45
shareholders' meetings 102–3
Shiba Electric Corporation 70, 71
Shibusawa, E. 10
Shiina, T. 221
Shimosato, H. 16, 17, 88
Shinto shrines 32, 92–3
ship as training centre 113–14
shortwave communication with guards 113
shrines 32, 92–3
Simon, H. A. 243
sincerity 45
smoke sensors 130
social education 17–18
social networks 214
social systems industry 116, 191, 204, 206, 212, 220, 239
socialization 17–18
societal benefit of business 10, 29, 33, 34, 208
societal perception and societal behaviour 217–18
Sogo Keibi Hosho 64–7
sokaiya 102
songs, company 84–5, 149–51
Sony Corporation 238
Sorensen, E. P. 6, 42–3
South Korea, SECOM activities in 169–76
SP (Security Patrol) Alarm System 71–2, 84, 89, 91, 95–6, 124, 126, 140, 225
speech processing research 210
sport, Iida's interest in 15–18
sprinkler systems 130
staff *see* employees
starting new business 36–7
statistical data 58
status quo 234
stealing by employees 77–80, 107
Stock Exchange listing of NKH 108–11
strikes 59, 60
study scholarships 118–21

success 239
succession to Iida 212
Sugimachi, Y. 153, 192
Sun Tribe 15
supervision 191
surrender of Japan 14, 235
system auditing 224
system design 37
System Design and System
   Development (SD) Centre,
   SECOM 145-6

Taiwan, SECOM activities in 158-
   68
Taiyozoku 15
Takahashi, H. 230-1
takeovers *see* acquisitions
Tao, Y. 215-20
Tarui, I. 122
technological innovation 68-85, 95,
   187, 214
Technomart 216
Tecno-Bio Company 230
telecommunications networks 91
telephone privacy 134
television dramatisation of security
   industry 63-4, 76
Telidon 215-16
temples 32, 91-3
terrorism 76-7, 105-6, 224
Thai SECOM Pitaki 194-7
Thailand, SECOM activities in 191-
   7
theft by employees 77-80, 107
thinking: logical 37; open 68
time as a resource 154
Toda, J. 26-9, 38, 44-5, 47, 61, 121,
   238
Tokubetsu Boei Hosho 99, 100
Tokugawa reign 8-9
Tokyo Broadcasting Station (TBS)
   63-4
Tokyo Olympic Games 53-7
Tomahawk Mach 132
Tomita, N. 158
total building service 139-42
total security packages 129
Totax-T 141
Toyota 237
tracing technology 204

trade unions 58-60
training 79, 111-21; costs of 94; in
   crisis response 83; in English 63,
   119; guards 63, 104-5, 106-7; in
   Japanese 210; managers 22;
   research staff 208; Taiwan 166-7;
   Thailand 195; USA 190; *see also*
   education
transport services 228
travel services 227
trend analysis 58
trout farms 128
trust 88, 154, 238
Tsuken (Denki Tsushin Kenkyujo)
   200-1
Tsushima, S. 196

uniforms of guards 104
unions 58-60
uniqueness 29
United States: research students
   from 209; SECOM activities in
   153, 154, 176-91
Ushio, J. 216
Usui, Y. 99-100

Valley Alarm 177, 179
value added networks (VANs) 214,
   220-5
vertical integration of NKH/
   SECOM 143-51
videotex 215-20
Videotex Centre 216-20
VIEWCOM 218-20
vision, entrepreneurial 2-3, 23, 212
voice scrambling research 210
voice synthesis research 210

Wada, H. 153
wages 58-9
Wakayama 137-8
war: World War II 12-14, 235
water sprinkler systems 130
Westec Security Corporation 177-9,
   181-4
western influences on Japan 236-7
Westinghouse Electric Corporation
   177-8
working environment 111
World War II 12-14, 235

yakuza 102
Yamakawa, K. 149
Yasu, H. 84
Yasukawa, D. 66
YKK 237

Yoshida, S. 65
Yoshida, T. 237

Zenkoku Bohan/Keibigyo Kyokai
Rengo-kai 39, 105

Due